◆ 山东省双语示范课程教材
◆ 山东省精品课程教材

# Functions of Complex Variables
# 复变函数论

## 第 二 版

Ma Lixin
马立新(编)

China Agriculture Press
中国农业出版社

#### 图书在版编目（CIP）数据

复变函数论 / 马立新编 . —2 版 . —北京：中国农业出版社，2014.12
ISBN 978-7-109-19726-8

Ⅰ.①复… Ⅱ.①马… Ⅲ.①复变函数-高等学校-教材 Ⅳ.①O174.5

中国版本图书馆 CIP 数据核字（2014）第 251917 号

中国农业出版社出版
（北京市朝阳区麦子店街 18 号楼）
（邮政编码 100125）
责任编辑 刘明昌

北京中科印刷有限公司印刷　新华书店北京发行所发行
2014 年 12 月第 2 版　2014 年 12 月第 2 版北京第 1 次印刷

开本：720mm×960mm　1/16　印张：14.5
字数：260 千字
定价：35.00 元
（凡本版图书出现印刷、装订错误，请向出版社发行部调换）

# 前 言

[复变函数论]

  伴随着我国高等教育改革形势的发展，高等教育的人才培养模式和教学方式以及教学方法正在发生重大变化。教育部于2001年在《关于加强高等学校本科教学工作提高教学质量的若干意见》（教高【2001】4号）中，明确要求高校要积极开展公共课和专业课教学双语教学的研究和实践。自2005年开始，德州学院采取一系列措施，开展这项创新教育改革活动，推出第一批英汉双语专业课程教学，《复变函数论》即是其中之一。对学生调查的结果及学校组织的听课评议结果均反映良好，其特色得到了山东省课程建设委员会专家的充分肯定，2009年被评为山东省首批双语示范课程、2012年被评为省级精品课程。

  对于双语教学，选一本合适的教材极其重要。由于教育体制不同，我国高校复变函数论教材与英美原版教材差别较大，很难找到完全适合我国开展《复变函数论》双语教学的英文原版教材。自2006年起，德州学院选用了 James Ward Brown & Ruel V. Churchill 编写的英文原版教材 Complex Variables and Applications（第七版），同时选用了该书配套的由邓冠铁等人翻译的中文版教材，在2005级、2006级、2007级进行了双语教学。经过3年的教学实践，笔者认为，这本教材内容涵盖较多、习题量较大，但选材和习题配置与现行中文教材差异较大。鉴于国内目前尚未见到同类英文教材，我们借鉴国外相关教材等英文原版编写了《复变函数论》双语教材并于 2009 年出版。经过试用，效果良好，故再次出版。第二版修正了原来的一些不当之处，并增加了习题解答提示。

  本教材共 6 章，主要内容包括复数与复变函数、解析函数、复变函数的积分、级数、留数及其应用和共形映射等，较全面、系统地介绍了复变函数的基础知识。内容处理上重点突出、叙述简明，每节末附有适量习题供读者选用，适合高等师范院校数学系及普通综合性大学数学系高年级学生使用。

  限于编者水平有限，书中不足之处，敬请读者批评指正。

编 者

2014 年 10 月

# Contents

[复变函数论]

## 前言

## Chapter Ⅰ  Complex Numbers and Functions

**1 Complex Numbers** ······································································ 1
  1.1  Complex Number Field ··········································· 1
  1.2  Complex Plane ························································ 2
  1.3  Modulus, Conjugation, Argument, Polar Representation ··· 3
  1.4  Powers and Roots of Complex Numbers ················ 6
  Exercises ············································································ 9
**2 Regions in the Complex Plane** ············································· 13
  2.1  Some Basic Concept ············································· 13
  2.2  Domain and Jordan Curve ································· 15
  Exercises ·········································································· 17
**3 Functions of a Complex Variable** ······································ 18
  3.1  The Concept of Functions of a Complex Variable ·············· 18
  3.2  Limits and Continuous ········································· 19
  Exercises ·········································································· 21
**4 The Extended Complex Plane and the Point at Infinity** ·········· 22
  4.1  The Spherical Representation, the Extended Complex Plane ········ 22
  4.2  Some Concepts in the Extended Complex Plane ············· 23
  Exercises ·········································································· 24

## Chapter Ⅱ  Analytic Functions

**1 The Concept of the Analytic Function** ·································· 25

    1.1  The Derivative of the Functions of a Complex Variable ......... 25
    1.2  Analytic Functions ......................................................... 29
    Exercises ............................................................................. 30
  2  **Cauchy-Riemann Equations** ............................................... 30
    Exercises ............................................................................. 35
  3  **Elementary Functions** ....................................................... 36
    3.1  The Exponential Function ............................................. 36
    3.2  Trigonometric Functions ............................................... 37
    3.3  Hyperbolic Functions .................................................... 39
    Exercises ............................................................................. 40
  4  **Multi-Valued Functions** .................................................... 42
    4.1  The Logarithmic Function ............................................. 42
    4.2  Complex Power Functions ............................................. 44
    4.3  Inverse Trigonometric and Hyperbolic Functions ............ 46
    Exercises ............................................................................. 48

## Chapter III  Complex Integration

  1  **The Concept of Contour Integrals** ..................................... 50
    1.1  Integral of a Complex Function over a Real Interval ....... 50
    1.2  Contour Integrals ........................................................... 51
    Exercises ............................................................................. 56
  2  **Cauchy-Goursat Theorem** ................................................. 60
    2.1  Cauchy Theorem ........................................................... 60
    2.2  Cauchy Integral Formula ............................................... 64
    2.3  Derivatives of Analytic Functions .................................. 66
    2.4  Liouville's Theorem and the Fundamental Theorem of Algebra ...... 67
    Exercises ............................................................................. 68
  3  **Harmonic Functions** ......................................................... 71
    Exercises ............................................................................. 73

## Chapter IV  Series

  1  **Basic Properties of Series** .................................................. 74
    1.1  Convergence of Sequences ............................................ 74

    1.2   Convergence of Series ………………………………………… 76

    1.3   Uniform convergence ………………………………………… 78

    **Exercises** ………………………………………………………………… 79

2  **Power Series** ………………………………………………………… 80

    **Exercises** ………………………………………………………………… 84

3  **Taylor Series** ………………………………………………………… 85

    **Exercises** ………………………………………………………………… 89

4  **Laurent Series** ……………………………………………………… 93

    **Exercises** ………………………………………………………………… 95

5  **Zeros of an Analytic Functions and Uniquely Determined Analytic Functions** ……………………………………………………………… 97

    5.1   Zeros of Analytic Functions ………………………………… 97

    5.2   Uniquely Determined Analytic Functions ………………… 101

    5.3   Maximum Modulus Principle ……………………………… 103

    **Exercises** ……………………………………………………………… 104

6  **The Three Types of Isolated Singular Points at a Finite Point** ……… 106

    **Exercises** ……………………………………………………………… 112

7  **The Three Types of Isolated Singular Points at a Infinite Point** …… 113

    **Exercises** ……………………………………………………………… 114

# Chapter Ⅴ  Calculus of Residues

1  **Residues** ……………………………………………………………… 115

    1.1   Residues ……………………………………………………… 115

    1.2   Cauchy's Residue Theorem ………………………………… 117

    1.3   The Calculus of Residue …………………………………… 119

    **Exercises** ……………………………………………………………… 126

2  **Applications of Residue** …………………………………………… 128

    2.1   The Type of Definite Integral $\int_0^{2\pi} F(\sin\theta, \cos\theta)\,d\theta$ ………… 128

    2.2   The Type of Improper Integral $\int_{-\infty}^{\infty} \frac{p(x)}{q(x)} dx$ …………………… 130

    2.3   The Type of Improper Integral $\int_{-\infty}^{+\infty} \frac{p(x)}{q(x)} \sin ax\,dx$ or

         $\int_{-\infty}^{+\infty} \frac{p(x)}{q(x)} \cos ax\,dx$ ……………………………………………… 132

  Exercises ............................................................ 133
**3 Argument Principle** ............................................. 136
  Exercises ............................................................ 138

# Chapter Ⅵ Conformal Mappings

**1 Analytic Transformation** ..................................... 140
  1.1 Preservation of Domains of Analytic Transformation ......... 140
  1.2 Conformality of Analytic Transformation ........................ 141
  Exercises ............................................................ 145
**2 Rational Functions** ............................................. 145
  2.1 Polynomials ................................................... 145
  2.2 Rational Functions .......................................... 146
  Exercises ............................................................ 148
**3 Fractional Linear Transformations** ........................ 149
  Exercises ............................................................ 154
**4 Elementary Conformal Mappings** ......................... 155
  Exercises ............................................................ 163
**5 The Riemann Mapping Theorem** .......................... 165
  Exercises ............................................................ 167

# Appendix

  Appendix 1 ........................................................ 169
  Appendix 2 ........................................................ 186
**Answers** ............................................................... 212
**Bibliography** ........................................................ 224

# Chapter I

## Complex Numbers and Functions

The complex function is a function of complex variables. The complex functions is a branch of analytics, it is also called Complex Analysis.

One of the advantages of dealing with the real numbers instead of the rational numbers is that certain equations which do not have any solutions in the rational numbers have a solution in the real numbers. For instance, $x^2=2$ or $x^2=3$ are such equations. However, we also know some equations having no solution in the real numbers, for instance, $x^2=-1$ or $x^2=-2$. In this chapter, we define a new kind of numbers where such equations have solutions. We will survey the algebraic and geometric structure of the complex number system.

## 1 Complex Numbers

### 1.1 Complex Number Field

**Definition 1.1.1** We call the numbers form $z=x+iy$ as complex numbers, in which $x$ and $y$ are all real numbers, $i$ is a number that satisfy $i^2=-1$ and $i$ is called imaginary unit. We call $x$ and $y$ the real and imaginary parts of $z$ and denote this by

$$\text{Re } z=x, \quad \text{Im } z=y \tag{1.1.1}$$

We notice that $z=x$ is a real numbers if $y=0$, and $z=iy$ is called pure imaginary number if $x=0$.

Two complex numbers $z_1=x_1+iy_1$ and $z_2=x_2+iy_2$ are equal if and only if they have the same real part and the same imaginary part.

The ordinary laws of arithmetic operations are defined as:

$$(x_1+iy_1)\pm(x_2+iy_2)=(x_1\pm x_2)+i(y_1\pm y_2)$$
$$(x_1+iy_1)(x_2+iy_2)=(x_1x_2-y_1y_2)+i(x_1y_2+x_2y_1)$$
$$\frac{(x_1+iy_1)}{(x_2+iy_2)}=\frac{x_1x_2+y_1y_2}{x_2^2+y_2^2}+i\frac{x_2y_1-x_1y_2}{x_2^2+y_2^2} \tag{1.1.2}$$

As a special case the reciprocal of a complex number $z=x+iy\neq 0$ is given by

$$\frac{1}{x+iy}=\frac{x}{x^2+y^2}+i\frac{-y}{x^2+y^2}$$

From the discussion above, we conclude that the set **C** of all complex numbers becomes a field, called the field of complex numbers, or the complex field. We may consider **R** as a subset of **C**.

## 1.2 Complex Plane

For mapping: $\mathbf{C}\to\mathbf{R}^2$: $z=x+iy\mapsto(x, y)$ then built a one-to-one correspondence between the set of complex numbers and the plane $\mathbf{R}^2$.

With respect to a given rectangular coordinate system in a plane, the complex number $z=x+iy$ can be represented by the point with coordinates $(x, y)$. The first coordinate axis ($x$-axis) takes the name of real axis, and the second coordinate axis ($y$-axis) is called the imaginary axis. The plane itself is referred to as the complex plane.

It is natural to associate any nonzero complex number $z=x+iy$ with the directed line segment, or vector, from the origin to the point $(x, y)$ that represents $z$ in the complex plane. In fact, we often refer to $z$ as the point $z$ or the vector $z$. The number, the point, and the vector will be denoted by the same letter $z$.

According to the definition of the sum of two complex numbers $z_1=x_1+iy_1$ and $z_2=x_2+iy_2$, the number $z_1+z_2$ corresponds to the point $(x_1+x_2, y_1+y_2)$. It also corresponds to a vector with those coordinates as its components. Hence $z_1+z_2$ may be obtained as shown in Fig. 1. The difference $z_1-z_2=z_1+(-z_2)$ corresponds to the sum of the vectors for $z_1$ and $-z_2$ (Fig. 2).

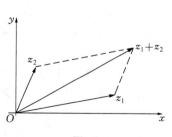

Fig. 1

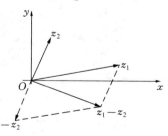

Fig. 2

Chapter I  Complex Numbers and Functions

## 1.3  Modulus, Conjugation, Argument, Polar Representation

**Definition 1.1.2**  If $z=x+iy$ then we define
$$|z|=\sqrt{x^2+y^2} \qquad (1.1.3)$$
to be the absolute value of $z$.

If we think of $z$ as a point in the plane $(x, y)$, then $|z|$ is the length of the line segment from the origin to $z$. It reduces to the usual absolute value in the real number system when $y=0$.

**Theorem 1.1.1**  The absolute value of a complex number satisfies the following properties. If $z_1$, $z_2$, $z$ are complex numbers, then
$$|-z|=|z| \qquad (1.1.4)$$
$$|z_1 z_2|=|z_1||z_2|, \qquad \left|\frac{z_1}{z_2}\right|=\frac{|z_1|}{|z_2|} \qquad (1.1.5)$$
$$|z_1 \pm z_2| \leqslant |z_1|+|z_2|, \quad ||z_1|-|z_2|| \leqslant |z_1 \pm z_2| \qquad (1.1.6)$$

(1.1.6) is called the triangle inequality because, if we represent $z_1$ and $z_2$ in the plane, (1.1.6) says that the length of one side of the triangle is less than the sum of the lengths of the other two sides. Or, the shortest distance between two points is a straight line.

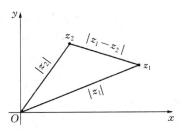

Fig. 3

By mathematical induction we also get:

**Theorem 1.1.2**  If $z_1$, $z_2$, $\cdots$, $z_n$ are complex numbers then we have
$$|z_1+z_2+\cdots+z_n| \leqslant |z_1|+|z_2|+\cdots+|z_n| \qquad (1.1.7)$$

**Definition 1.1.3**  The complex conjugate, or simply the conjugate, of a complex number $z=x+iy$ is defined as the complex number $x-iy$ and is denoted by $\bar{z}$; that is $\bar{z}=x-iy$.

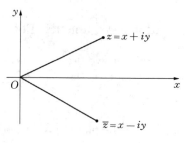

Fig. 4

The point $z$ and its conjugate $\bar{z}$ lie symmetrically with respect to the real axis. This is also easy; in fact, is the point obtained by reflecting $z$ across the $x$-axis (i. e. , the real axis). A number is real if and only if it is equal to its conjugate.

**Theorem 1. 1. 3** The complex conjugate of a complex number satisfies the following properties. If $z_1$, $z_2$, $z$ are complex numbers, then

$$\mathrm{Re}z = \frac{z+\bar{z}}{2}, \quad \mathrm{Im}z = \frac{z-\bar{z}}{2i} \tag{1.1.8}$$

$$z\bar{z} = |z|^2, \quad |z| = |\bar{z}|, \quad \bar{\bar{z}} = z \tag{1.1.9}$$

$$\overline{z_1 + z_2} = \bar{z}_1 + \bar{z}_2, \quad \overline{z_1 - z_2} = \bar{z}_1 - \bar{z}_2 \tag{1.1.10}$$

$$\overline{z_1 z_2} = \bar{z}_1 \bar{z}_2, \quad \overline{\left(\frac{z_1}{z_2}\right)} = \frac{\bar{z}_1}{\bar{z}_2} \; (z_2 \neq 0) \tag{1.1.11}$$

Let $(x, y) = x + iy$ be a complex number. We know that any point in the plane can be represented by polar coordinates $(r, \theta)$:

$$x = r\cos\theta \quad y = r\sin\theta \tag{1.1.12}$$

Hence we can write $z = (x, y) = x + iy = r(\cos\theta + i\sin\theta)$. In this trigonometric form of a complex number $r$ is always $\geq 0$ and equal to the modulus $|z|$.

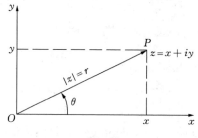

Fig. 5

**Definition 1. 1. 4** The polar angle $\theta$ is called the argument of the complex

Chapter I  Complex Numbers and Functions

number, and we denote it by Argz. The principal value of Argz, denoted by argz, is that unique value $\theta$ such that $-\pi < \theta \leqslant \pi$. Note that
$$\text{Arg}z = \{\arg z + 2n\pi : n = 0, \pm 1, \pm 2, \cdots\}$$
Simply, we write
$$\text{Arg}z = \arg z + 2n\pi \quad (n=0, \pm 1, \pm 2, \cdots) \tag{1.1.13}$$
Also, when $z$ is a negative real number, argz has value $\pi$, not $-\pi$.

$$\arg z \atop (z \neq 0) = \begin{cases} \arctan \dfrac{y}{x} \cdots\cdots\cdots (x > 0) \\[4pt] \dfrac{\pi}{2} \cdots\cdots\cdots\cdots (x = 0, y > 0) \\[4pt] \arctan \dfrac{y}{x} + \pi \cdots\cdots (x < 0, y \geqslant 0) \\[4pt] \arctan \dfrac{y}{x} - \pi \cdots\cdots (x < 0, y < 0) \\[4pt] -\dfrac{\pi}{2} \cdots\cdots\cdots\cdots (x = 0, y < 0) \end{cases} \tag{1.1.14}$$

where $-\dfrac{\pi}{2} < \arctan \dfrac{y}{x} < \dfrac{\pi}{2}$.

We list some important identity involving arguments:
$$\text{Arg}(z_1 z_2) = \text{Arg} z_1 + \text{Arg} z_2 \tag{1.1.15}$$
$$\text{Arg}(z_2^{-1}) = -\text{Arg} z_2 \tag{1.1.16}$$
$$\text{Arg}\left(\dfrac{z_1}{z_2}\right) = \text{Arg} z_1 - \text{Arg} z_2 \tag{1.1.17}$$

**Example 1**  Compute $\text{Arg}(2-2i)$ and $\text{Arg}(-3+4i)$.
$$\text{Arg}(2-2i) = \arg(2-2i) + 2n\pi = \arctan \dfrac{-2}{2} + 2n\pi = -\dfrac{\pi}{4} + 2n\pi$$
$$\text{Arg}(-3+4i) = \arg(-3+4i) + 2n\pi = \arctan \dfrac{4}{-3} + \pi + 2n\pi = (2n+1)\pi - \arctan \dfrac{4}{3}$$

**Example 2**  To find the principal argument argz when
$$z = \dfrac{-2}{1+\sqrt{3}i}$$
Observe that
$$\text{Arg}z = \text{Arg}(-2) - \text{Arg}(1+\sqrt{3}i)$$
Since

$$\arg(-2) = \pi, \arg(1+\sqrt{3}i) = \frac{\pi}{3}$$

One value of $\text{Arg}z$ is $\frac{2\pi}{3}$; and, because $\frac{2\pi}{3}$ is between $-\pi$ and $\pi$, we find that $\arg z = \frac{2\pi}{3}$.

## 1.4 Powers and Roots of Complex Numbers

(1) Powers

**Definition 1.1.5** We define the expression $e^{i\theta}$ to be

$$e^{i\theta} = \cos\theta + i\sin\theta \text{ (Euler's formula)} \quad (1.1.18)$$

where $\theta$ is to be measured in radians. Thus $e^{i\theta}$ is a complex number.

It enables us to write the polar form of a complex number in exponential form as

$$z = re^{i\theta}$$

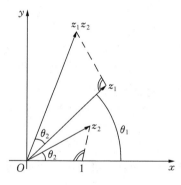

Fig. 6

**Theorem 1.1.4**
$$e^{i\theta_1} e^{i\theta_2} = e^{i(\theta_1+\theta_2)} \quad (1.1.19)$$
$$z_1 z_2 = r_1 r_2 e^{i(\theta_1+\theta_2)} \quad (1.1.20)$$
$$\frac{z_1}{z_2} = \frac{r_1}{r_2} e^{i(\theta_1-\theta_2)} \quad (1.1.21)$$
$$z^{-1} = \frac{1}{r} e^{-i\theta} \quad (1.1.22)$$

Expressions (1.1.19), (1.1.20), (1.1.21), and (1.1.22) are, of course, easily remembered by applying the usual algebraic rules for real numbers and $e^x$.

# Chapter I  Complex Numbers and Functions

**Definition 1.1.6**  We define the powers of $z=re^{i\theta}$ is
$$z^n = r^n e^{in\theta} \quad (n=0, \pm 1, \pm 2, \cdots) \tag{1.1.23}$$
For $r=1$ we obtain de Moivre's formula
$$(\cos\theta + i\sin\theta)^n = \cos n\theta + i\sin n\theta \quad (n=0, \pm 1, \pm 2, \cdots) \tag{1.1.24}$$
Which provides an extremely simple way to express $\cos n\theta$ and $\sin n\theta$ in terms of $\cos\theta$ and $\sin\theta$.

**Example 3**  The number $-1-i$ has exponential form
$$-1-i = \sqrt{2}\, e^{i(-\frac{3}{4}\pi)} \tag{1.1.25}$$

Expression (1.1.25) is only one of an infinite number of possibilities for the exponential from of $-1-i$:
$$-1-i = \sqrt{2}\, e^{i(-\frac{3}{4}\pi + 2n\pi)} \quad (n=0, \pm 1, \pm 2, \cdots)$$

**Example 4**  Put $(\sqrt{3}+i)^7$ in rectangular form.

We write
$$(\sqrt{3}+i)^7 = (2e^{i\pi/6})^7 = 2^7 e^{i7\pi/6} = (2^6 e^{i\pi})(2e^{i\pi/6}) = -64(\sqrt{3}+i)$$

(2) Square Roots

To find the $n$ th root of a complex number $z$ we have to solve the equation
$$w^n = z \tag{1.1.26}$$
Suppose that $z \neq 0$, $z = re^{i\theta}$, $w = \rho e^{i\phi}$. Then (1.1.26) takes the form
$$\rho^n e^{in\phi} = re^{i\theta}$$
This equation is certainly fulfilled if $\rho^n = r$ and $n\phi = \theta + 2k\pi$. Hence we obtain the root
$$w = \sqrt[n]{r}\, e^{i\frac{\theta + 2k\pi}{n}}, \quad k=0, \pm 1, \pm 2, \cdots$$
However, only the values $k=0, 1, 2, \cdots, n-1$ give different value of $z$.

**Definition 1.1.7**  We define the $n$ th root of a complex number $z$ is
$$w = \sqrt[n]{r}\, e^{i\frac{\theta + 2k\pi}{n}}, \quad k=0, 1, 2, \cdots, n-1 \tag{1.1.27}$$
There are $n$ th roots of any complex number $z \neq 0$. They have the same modulus, and their arguments are equally spaced.

Geometrically, the $n$ th roots are the vertices of a regular polygon with $n$ sides.

**Example 5**  Determine the $n$ th roots of unity.

We write
$$1 = 1e^{i0}$$
and find that

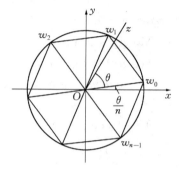

Fig. 7

$$1^{\frac{1}{n}} = \sqrt[n]{1}\, e^{i\frac{0+2k\pi}{n}} = e^{i\frac{2k\pi}{n}}\ (k=0,\ 1,\ 2,\ \cdots,\ n-1)$$

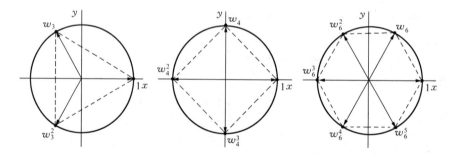

Fig. 8

**Example 6** Find the value of $\sqrt[4]{1+i}$.

Because $1+i = \sqrt{2}\, e^{i\frac{\pi}{4}}$. So that $\sqrt[4]{1+i} = \sqrt[8]{2}\, e^{i\frac{\frac{\pi}{4}+2k\pi}{4}}\ (k=0,\ 1,\ 2,\ 3)$.

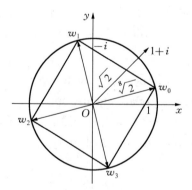

Fig. 9

Chapter I  Complex Numbers and Functions

## Exercises

1. Verify that
   (a) $(\sqrt{2}-i)-i(1-\sqrt{2}i) = -2i$  (b) $(2,-3)(-2,1) = (-1,8)$
   (c) $(3,1)(3,-1)\left(\frac{1}{5}, \frac{1}{10}\right) = (2,1)$

2. If $z = x+iy$ ($x$ and $y$ real), find the real and imaginary parts of
   (a) $z^4$   (b) $\frac{1}{z}$   (c) $\frac{z-1}{z+1}$   (d) $\frac{1}{z^2}$

3. Express the following complex numbers in the form $x+iy$, where $x$ and $y$ are real numbers.
   (a) $(-1+3i)^{-1}$   (b) $(1+i)(2-i)$   (c) $(2i+1)\pi i$
   (d) $(i+1)(i-2)(i+3)$   (e) $(1+i)^{-1}$   (f) $\frac{1}{3+i}$
   (g) $\frac{2+i}{2-i}$   (h) $\frac{1}{-1+i}$

4. Show that
   (a) $\operatorname{Re}(iz) = -\operatorname{Im}z$   (b) $\operatorname{Im}(iz) = \operatorname{Re}z$

5. Find the values of
   (a) $(1+2i)^3$   (b) $\frac{5}{-3+4i}$   (c) $\left(\frac{2+i}{3-2i}\right)^2$
   (d) $(1+i)^n + (1-i)^n$

6. Reduce each of these quantities to a real number.
   (a) $\frac{1+2i}{3-4i} + \frac{2-i}{5i}$   (b) $\frac{5i}{(1-i)(2-i)(3-i)}$   (c) $(1-i)^4$

7. Show that
   (a) $(-1)z = -z$   (b) $\frac{1}{1/z} = z \quad (z \neq 0)$

8. Show that
$$\left(\frac{-1 \pm i\sqrt{3}}{2}\right)^3 = 1 \quad \text{and} \quad \left(\frac{-1 \pm i\sqrt{3}}{2}\right)^6 = 1$$
for all combinations of signs.

9. Solve the quadratic equation
$$z^2 + (\alpha + i\beta)z + \gamma + i\delta = 0$$

10. Locate the numbers $z_1 + z_2$ and $z_1 - z_2$ vectorially when

(a) $z_1 = 2i, z_2 = \dfrac{2}{3} - i$  (b) $z_1 = (-\sqrt{3}, 1), z_2 = (-\sqrt{3}, 0)$

(c) $z_1 = (-3, 1), z_2 = (1, 4)$  (d) $z_1 = x_1 + iy_1, z_2 = x_1 - iy_1$

11. Verify that $\sqrt{2} \, |z| \geq |\text{Re} z| + |\text{Im} z|$.

12. Find the absolute value and conjugate of each of the following:

(a) $-2i(3+i)(2+4i)(1+i)$  (b) $\dfrac{(3+4i)(-1+2i)}{(-1-i)(3-i)}$

(c) $i^{17}$  (d) $(1+i)^6$

13. If $z$ and $w$ are complex numbers. To prove the following equations:

(a) $|z+w|^2 = |z|^2 + 2\text{Re} z\overline{w} + |w|^2$

(b) $|z-w|^2 = |z|^2 - 2\text{Re} z\overline{w} + |w|^2$

(c) $|z+w|^2 + |z-w|^2 = 2|z|^2 + 2|w|^2$

14. To show that

(a) $\overline{z+3i} = z - 3i$  (b) $\overline{iz} = -i\overline{z}$  (c) $\overline{(2+i)^2} = 3 - 4i$

15. To show that

(a) $\overline{z_1 z_2 z_3} = \overline{z_1}\,\overline{z_2}\,\overline{z_3}$  (b) $\overline{z^4} = \overline{z}^4$

16. To show that when $z_2$ and $z_3$ are nonzero

(a) $\overline{\left(\dfrac{z_1}{z_2 z_3}\right)} = \dfrac{\overline{z_1}}{\overline{z_2}\,\overline{z_3}}$  (b) $\left|\dfrac{z_1}{z_2 z_3}\right| = \dfrac{|z_1|}{|z_2||z_3|}$

17. To show that when $|z_3| \neq |z_4|$,

$$\left|\dfrac{z_1+z_2}{z_3+z_4}\right| \leq \dfrac{|z_1|+|z_2|}{||z_3|-|z_4||}$$

18. Show that

$$|\text{Re}(2+\overline{z}+z^3)| \leq 4 \text{ when } |z| \leq 1$$

19. To show that if $z$ lies on the circle $|z|=2$, then

$$\left|\dfrac{1}{z^4-4z^2+3}\right| \leq \dfrac{1}{3}$$

20. Prove that

(a) $z$ is real if and only if $\overline{z} = z$

(b) $z$ is either real or pure imaginary if and only if $\overline{z}^2 = z^2$

21. Use mathematical induction to show that when $n = 2, 3, \cdots$

(a) $\overline{z_1 + z_2 + \cdots + z_n} = \overline{z_1} + \overline{z_2} + \cdots + \overline{z_n}$

(b) $\overline{z_1 z_2 \cdots z_n} = \overline{z_1}\,\overline{z_2}\cdots\overline{z_n}$

## Chapter I  Complex Numbers and Functions

22. Let $a_0$, $a_1$, $a_2$, $\cdots$, $a_n$ $(n \geq 1)$ denote real numbers, and let $z$ be any complex number.

    To show that
    $$\overline{a_0 + a_1 z + a_2 z^2 + \cdots + a_n z^n} = a_0 + a_1 \bar{z} + a_2 \bar{z}^2 + \cdots + a_n \bar{z}^n$$

23. Show that the equation $|z - z_0| = R$ of a circle, centered at $z_0$ with radius $R$, can be written
    $$|z|^2 - 2\mathrm{Re}(z \overline{z_0}) + |z_0|^2 = R^2$$

24. Follow the steps below to give an algebraic derivation of the triangle inequality
    $$|z_1 + z_2| \leq |z_1| + |z_2|$$
    (a) Show that
    $$|z_1 + z_2|^2 = (z_1 + z_2)(\overline{z_1} + \overline{z_2}) = z_1 \overline{z_1} + (z_1 \overline{z_2} + \overline{z_1} z_2) + z_2 \overline{z_2}$$
    (b) Point out why
    $$z_1 \overline{z_2} + \overline{z_1} z_2 = 2\mathrm{Re}(z_1 \overline{z_2}) \leq 2 |z_1| \, |z_2|$$
    (c) Use the results in parts (a) and (b) to obtain the inequality
    $$|z_1 + z_2|^2 \leq (|z_1| + |z_2|)^2$$
    and note how the triangle inequality follows.

25. Let $R(z)$ be a rational function of $z$. Show that $\overline{R(z)} = R(\bar{z})$ if all the coefficients in $R(z)$ are real.

26. Verify by calculation that the values of
    $$\frac{z}{1 + z^2}$$
    for $z = x + iy$ and $z = x - iy$ are conjugate.

27. Prove that
    $$\left| \frac{a - b}{1 - \bar{a} b} \right| < 1$$
    if $|a| < 1$ and $|b| < 1$.

28. Prove that
    $$\left| \frac{a - b}{1 - \bar{a} b} \right| = 1$$
    if either $|a| = 1$ or $|b| = 1$. What exception must be made if $|a| = |b| = 1$?

## Functions of Complex Variables

29. Find the conditions under which the equation
$$az + b\bar{z} + c = 0$$
in one complex unknown has exactly one solution, and compute that solution.

30. If $|z_i| < 1$, $\lambda_i \geq 0$ for $i = 1, 2, \cdots, n$ and $\lambda_1 + \lambda_2 + \cdots + \lambda_n = 1$, show that
$$|\lambda_1 z_1 + \cdots + \lambda_n z_n| < 1$$

31. Show that there are complex numbers $z$ satisfying
$$|z - a| + |z + a| = 2|c|$$
if and only if $|a| \leq |c|$. If this condition is fulfilled, what are the smallest and largest values of $|z|$?

32. Find the principal argument arg$z$ when

    (a) $z = \dfrac{i}{-2 - 2i}$      (b) $z = (\sqrt{3} - i)^6$

33. Put the following complex numbers in polar form.

    (a) $1 + i$    (b) $1 + i\sqrt{2}$    (c) $-3$    (d) $4i$

    (e) $1 - i\sqrt{2}$    (f) $-5i$    (g) $-1 - i$

34. Put the following complex numbers in the ordinary form $x + iy$.

    (a) $e^{3i\pi}$    (b) $3e^{i\frac{\pi}{4}}$    (c) $\pi e^{-i\frac{\pi}{3}}$

    (d) $e^{-i\pi}$    (e) $e^{-i\frac{5\pi}{4}}$

35. Show that

    (a) $|e^{i\theta}| = 1$    (b) $\overline{e^{i\theta}} = e^{-i\theta}$

36. Use mathematical induction to show that
$$e^{i\theta_1} e^{i\theta_2} \cdots e^{i\theta_n} = e^{i(\theta_1 + \theta_2 + \cdots + \theta_n)} \quad (n = 2, 3, \cdots)$$

37. Use de Moivre's formula to derive the following trigonometric identities:

    (a) $\cos 3\theta = \cos^3 \theta - 3\cos\theta \sin^2 \theta$

    (b) $\sin 3\theta = 3\cos^2 \theta \sin\theta - \sin^3 \theta$

38. Express $\cos 4\theta$ and $\sin 5\theta$ in terms of $\cos\theta$ and $\sin\theta$.

39. Simplify $1 + \cos\theta + \cos 2\theta + \cdots + \cos n\theta$ and $1 + \sin\theta + \sin 2\theta + \cdots + \sin n\theta$.

40. To show that

    (a) $i(1 - \sqrt{3}i)(\sqrt{3} + i) = 2(1 + \sqrt{3}i)$     (b) $5i/(2 + i) = 1 + 2i$

(c) $(-1+i)^7 = -8(1+i)$  (d) $(1+\sqrt{3}i)^{-10} = 2^{-11}(-1+\sqrt{3}i)$

41. Find the roots of

    (a) $\sqrt{2i}$   (b) $\sqrt{1-\sqrt{3}i}$   (c) $\sqrt{i}$   (d) $\sqrt{-i}$

    (e) $\sqrt{1+i}$   (f) $\sqrt{\dfrac{1-i\sqrt{3}}{2}}$   (g) $\sqrt[4]{-1}$   (h) $\sqrt[4]{i}$

    (i) $(-16)^{1/4}$   (j) $(-8-8\sqrt{3}i)^{1/4}$

42. Let $z = \cos\dfrac{2\pi}{n}$ for an integer $n > 2$. Show that $1+z+z^2+\cdots+z^{n-1}=0$.

43. Prove that for any complex number $z \neq 1$ we have
$$1+z+z^2+\cdots+z^{n-1} = \frac{z^{n+1}-1}{z-1}$$

44. If $z \in C$ and $\mathrm{Re}(z^n) \geq 0$ for every positive integer $n$, show that $z$ is a positive real number.

45. When does $az + b\bar{z} + c = 0$ represent a line?

46. Write the equation of an ellipse, hyperbola, and parabola in complex form.

# 2 Regions in the Complex Plane

## 2.1 Some Basic Concept

**Definition 1.2.1** An $\varepsilon$-neighborhood
$$N(z_0, \varepsilon) = \{z : |z-z_0| < \varepsilon\} \tag{1.2.1}$$
of a given pint $z_0$. It consists of all points $z$ lying inside but not on a circle centered at $z_0$ and with a specified positive radius $\varepsilon$.

In other words, a neighborhood of $z_0$ is a set which contains all points sufficiently near to $z_0$.

**Definition 1.2.2** A deleted neighborhood, or centerless neighborhood,
$$N^0(z_0, \varepsilon) = \{z : 0 < |z-z_0| < \varepsilon\} \tag{1.2.2}$$
consisting of all points $z$ in an $\varepsilon$ neighborhood of $z_0$ except for the point $z_0$ itself.

**Definition 1.2.3** A point $z_0$ is said to be an interior point of a set $S$ whenever there is some neighborhood of $z_0$ that contains only points of $S$; it is called an exterior point of $S$ when there exists a neighborhood of it containing no

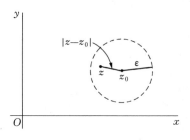

Fig. 10

points of S.

**Definition 1.2.4** The set of all interior points of a set $S$ is called the interior of $S$ and denoted by $S^0$ or Int $S$.

**Theorem 1.2.1** The interior of a set $S$ is the largest open set contained in $S$.

**Definition 1.2.5** The closure of $S$ is the smallest closed set which contains $S$. The closure is usually denoted by $\overline{S}$.

**Theorem 1.2.2** A point belongs to the closure of $S$ if and only if all its neighborhoods intersect $S$.

**Definition 1.2.6** If $z_0$ is neither an interior point nor an exterior point of $S$, it is a boundary point of $S$.

**Definition 1.2.7** The set of all boundary points of a set $S$ is called the boundary of $S$ and denoted by $\partial S$.

A boundary point is, therefore, a point all of whose neighborhoods contain points in $S$ and points not in $S$.

**Theorem 1.2.3** The boundary of $S$ is the closure minus the interior.

**Definition 1.2.8** A point $z_0$ is said to be an accumulation point of a set $S$ if each deleted neighborhood of $z_0$ contains at least one point of $S$.

**Theorem 1.2.4** A point $z_0$ is an accumulation point of a set $S$ if $S \cap N^0(z_0, \delta) \neq \Phi$ for all positive number $\delta$.

**Theorem 1.2.5** A point $z_0$ is an accumulation point is a point of $S$ if and only if every neighborhood of $z_0$ contains infinitely many points from $S$.

**Definition 1.2.9** If a set $S$ is closed, then it contains each of its accumulation points.

It follows that if accumulation point $z_0$ was not in $S$, it would be a boundary point of $S$.

**Theorem 1.2.6**　A closed set contains all of its boundary points.

**Definition 1.2.10**　A set of complex numbers $S$ to be compact if every sequence of elements of $S$ has a point of accumulation in $S$.

This property is equivalent to the following properties, which could be taken as alternate definitions:

**Theorem 1.2.7**

(1) Every infinite subset of $S$ has a point of accumulation in $S$.

(2) Every sequence of elements of $S$ has a convergent subsequence whose limit is in $S$.

**Definition 1.2.11**　A point $z_0$ is an isolated point of $S$ if $z_0$ has a neighborhood whose intersection with $S$ reduces to the point $z_0$.

## 2.2　Domain and Jordan Curve

**Definition 1.2.12**　A set is said to be open if it contains none of its boundary points.

**Theorem 1.2.8**　A set is open if it is a neighborhood of each of its elements.

The definition and theorem is interpreted to mean that the empty set is open (the condition is fulfilled because the set has no elements).

**Definition 1.2.13**　A nonempty open set $S$ is called connected if each pair of points $z_1$ and $z_2$ in it can be joined by a polygonal line, consisting of a finite number of line segments joined end to end, that lies entirely in $S$.

**Definition 1.2.14**　A nonempty connected open set is called a domain.

**Definition 1.2.15**　A domain together with some, none, or all of its boundary points is referred to as a region.

**Definition 1.2.16**　The closure of a region is called a closed region.

**Definition 1.2.17**　A set $C$ of points $z=(x, y)$ in the complex plane is said to be an arc if there exist continuous functions $x$ and $y$ of the real parameter $t$ on an interval $[a, b]$ such that

$$z(t) = x(t) + iy(t) \qquad (1.2.3)$$

**Definition 1.2.18**　An arc is simple, or a Jordan arc, if it does not cross itself; that is, for $a \leqslant t_1 \leqslant t_2 \leqslant b$, we have

$$z(t_1) = z(t_2) \Rightarrow t_1 = t_2, \text{or } t_1 = a \text{ and } t_2 = b$$

When the arc $C$ is simple such that $z(b)=z(a)$, we say that $C$ is a simple closed curve, or a Jordan curve.

**Definition 1.2.19**  If the derivative

$$z'(t) = x'(t) + iy'(t) \qquad (1.2.4)$$

exists and is $\neq 0$, the arc $C$ has a tangent whose direction is determined by $\arg z'(t)$. We shall say that the arc is differentiable if $z'(t)$ exists and is continuous; if, in addition, $z'(t) \neq 0$ the arc is said to be regular.

**Definition 1.2.20**  An arc is piecewise smooth arc, is an arc consisting of a finite number of smooth arcs joined end to end. Hence if equation (1.2.3) represents a contour, then $z(t)$ is continuous, whereas its derivative $z'(t)$ is piecewise continuous.

**Definition 1.2.21**  When only the initial and final values of $z(t)$ are the same, the contour $C$ is called a simple closed contour.

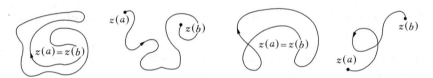

**Fig. 11**

**Example 7**  The unit circle

$$z = e^{i\theta} \ (0 \leqslant \theta \leqslant 2\pi)$$

about the origin is a simple closed curve, oriented in the counterclockwise direction. So is the circle

$$z = z_0 + Re^{i\theta} \ (0 \leqslant \theta \leqslant 2\pi)$$

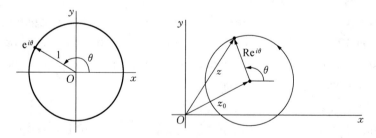

**Fig. 12**

centered at the point $z_0$ and with radius $R$.

**Theorem 1.2.9 (Jordan Curve Theorem)** Every Jordan curve in the complex plane determines exactly two regions.

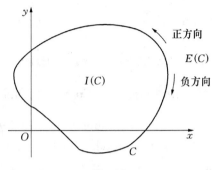

Fig. 13

## Exercises

1. Sketch the sets given by the following inequalities, respectively, and determine which are domains:
   (a) $|z-2+i|\leqslant 1$  (b) $|2z+3|>4$  (c) $|z-1+i|=1$
   (d) $|z+i|\leqslant 3$  (e) $|z-4i|\geqslant 4$  (f) $\mathrm{Re}(\bar{z}-i)=2$
   (g) $|2z-i|=4$  (h) $\mathrm{Im}z\geqslant 0$  (i) $\mathrm{Im}z>1$
   (j) $\mathrm{Im}z=1$  (k) $0\leqslant\mathrm{arg}z\leqslant\pi/4(z\neq 0)$  (l) $|z-4|\geqslant|z|$

2. Sketch the following curves with $0\leqslant t\leqslant 1$.
   (a) $\gamma(t)=1+it$  (b) $\gamma(t)=e^{-\pi ti}$
   (c) $\gamma(t)=e^{\pi ti}$  (d) $\gamma(t)=1+it+t^2$

3. Show by strict application of the definition that the closure of $|z-z_0|<\delta$ is $|z-z_0|\leqslant\delta$.

4. If $S$ is the set of complex numbers whose real and imaginary parts are rational, what is $\mathrm{Int}S$, $\bar{S}$, $\partial S$?

5. Show that the union of two regions is a region if and only if they have a common point.

6. Prove that the closure of a connected set is connected.

7. (a) Every infinite subset of $S$ has a point of accumulation in $S$.
   (b) Every sequence of elements of $S$ has a convergent subsequence whose

limit is in $S$.
8. Show that the Heine-Borel property can also be expressed in the following manner: Every collection of closed sets with an empty intersection contains a finite subcollection with empty intersection.
9. Show that the accumulation points of any set form a closed set.

# 3  Functions of a Complex Variable

## 3.1  The Concept of Functions of a Complex Variable

**Definition 1.3.1**  Let $E$ be a set of complex numbers. A function $f(z)$ defined on $E$ is a rule that assigns to each $z$ in $E$ a complex number $w$. The number $w$ is called the value of $f(z)$ at $z$ and is denoted by $f(z)$; that is $w = f(z)$. The set $E$ is called the domain of definition of $f(z)$.

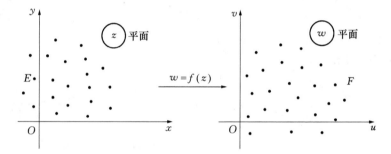

Fig. 14

We can write
$$f(z) = u(z) + iv(z)$$
where $u(z)$ and $v(z)$ are real numbers, and thus are real valued functions. We call $u(z)$ the real part of $f(z)$, and $v(z)$ the imaginary part of $f(z)$.

We shall usually write
$$z = x + iy$$
where $x$, $y$ are real. Then the values of the function $f(z)$ can be written in the form
$$f(z) = f(x + iy) = u(x, y) + v(x, y)i$$
viewing $u$, $v$ as functions of the two real variables $x$ and $y$.

**Example 8** For the function
$$f(z) = x^3 y + i\sin(x+y)$$
we have the real part
$$u(x,y) = x^3 y$$
and the imaginary part
$$v(x,y) = \sin(x+y)$$

## 3.2 Limits and Continuous

**Definition 1.3.2** Let $f(z)$ be a function on $E$, and a point $z_0$ in $E$. Let $w_0$ be a complex number. The function $f(z)$ is said to have the limit $w_0$ as $z$ tends to $z_0$,
$$\lim_{z \to z_0} f(z) = w_0 \qquad (1.3.1)$$
if and only if the following condition is satisfied: for each positive number $\varepsilon$, there is a positive number $\delta$ such that
$$|f(z) - w_0| < \varepsilon \text{ whenever } z \in E \text{ and } 0 < |z - z_0| < \delta \qquad (1.3.2)$$

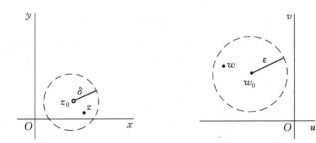

Fig. 15

If $z_0$ is an interior point of the domain of definition of $f(z)$, and limit (1.3.1) is to exist, the first of inequalities (1.3.2) must hold for all points in the deleted neighborhood $N^0(z_0, \delta)$. Thus the symbol $z \to z_0$ implies that $z$ is allowed to approach $z_0$ in an arbitrary manner, not just from some particular direction.

This definition makes use of the absolute value. Since the notion of absolute value has a meaning for complex as well as for real numbers, we can use the same definition regardless of whether the variable $z$ and the function $f(z)$ are real or complex.

We can establish a connection between limits of functions of a complex variable and limits of real-valued functions of two real variables.

**Theorem 1.3.1** Suppose that $f: E \to C$, $z_0 = x_0 + iy_0 \in E^0$, $w_0 = u_0 + iv_0$ and
$$f(z) = u(x,y) + iv(x,y)$$
Then
$$\lim_{z \to z_0} f(z) = w_0 \tag{1.3.3}$$
if and only if
$$\lim_{(x,y) \to (x_0,y_0)} u(x,y) = u_0 \quad \text{and} \quad \lim_{(x,y) \to (x_0,y_0)} v(x,y) = v_0 \tag{1.3.4}$$

The proof is easy, only use the definition of the limit.

**Definition 1.3.3** A function $f(z)$ is said to be continuous at a point $z_0$ if and only if
$$\lim_{z \to z_0} f(z) = f(z_0) \tag{1.3.5}$$

Statement (1.3.5) satisfy if and only if that, for each positive number $\varepsilon$, there is a positive number $\delta$, whenever $|z - z_0| < \delta$ such that
$$|f(z) - f(z_0)| < \varepsilon \tag{1.3.6}$$

A continuous function is continuous at all points where it is defined.

**Theorem 1.3.2** Suppose that $f: E \to C$, $z_0 = x_0 + iy_0 \in E^0$, $w_0 = u_0 + iv_0$ and
$$f(z) = u(x,y) + iv(x,y)$$
Then
$$\lim_{z \to z_0} f(z) = f(z_0) \tag{1.3.7}$$
if and only if
$$\lim_{(x,y) \to (x_0,y_0)} u(x,y) = u(x_0,y_0) \quad \text{and} \quad \lim_{(x,y) \to (x_0,y_0)} v(x,y) = v(x_0,y_0)$$
$$\tag{1.3.8}$$

These definitions are completely analogous to those which you have had in analysis, so we do not spend much time on them. As usual, we have the rules for limits and continuous of sums, products, quotients as in calculus.

**Theorem 1.3.3** Let $f(z)$, $g(z)$ be continuous function whose domain of definition contains a neighborhood of a point $z$. Then $f(z) + g(z)$ and $f(z)g(z)$ are both continuous. Also, $f(z)/g(z)$ is continuous provided

$g(z) \neq 0$.

**Theorem 1.3.4** Let $w = f(z)$. Assume that $f(z)$ is continuous at $z$, and $g(w)$ is continuous at $w$. Then the composite function $g(f(z))$ is continuous at $z$.

## Exercises

1. For each of the functions below, describe the domain of definition that is understood:

   (a) $f(z) = \dfrac{1}{z^2 + 1}$
   
   (b) $f(z) = \arg\left(\dfrac{1}{z}\right)$
   
   (c) $f(z) = \dfrac{z}{z + \bar{z}}$
   
   (d) $f(z) = \dfrac{1}{1 - |z|^2}$

2. Write the function $f(z) = z^3 + z + 1$ in the form $f(z) = u(x, y) + iv(x, y)$.

3. Write the function
$$f(z) = z + \frac{1}{z} \quad (z \neq 0)$$
in the form $f(z) = u(r, \theta) + iv(r, \theta)$.

4. Suppose that $f(z) = x^2 - y^2 - 2y + i(2x - 2xy)$, where $z = x + iy$. Use the expressions
$$x = \frac{z + \bar{z}}{2}, \quad y = \frac{z - \bar{z}}{2i}$$
to write $f(z)$ in terms of $z$, and simplify the result.

5. Use definition of limit to prove that

   (a) $\lim\limits_{z \to z_0} \operatorname{Re} z = \operatorname{Re} z_0$
   
   (b) $\lim\limits_{z \to z_0} \bar{z} = \bar{z_0}$
   
   (c) $\lim\limits_{z \to 0} \dfrac{\bar{z}^2}{z} = 0$

6. Let $a$, $b$, and $c$ denote complex constants. Then use definition of limit to show that

   (a) $\lim\limits_{z \to z_0}(az + b) = az_0 + b$
   
   (b) $\lim\limits_{z \to z_0}(z^2 + c) = z_0^2 + c$
   
   (c) $\lim\limits_{z \to 1 - i}[x + i(2x + y)] = 1 + i(z = x + iy)$

7. Let $n$ be a positive integer and let $P(z)$ and $Q(z)$ be polynomials, where $Q(z_0) \neq 0$. To find

   (a) $\lim\limits_{z \to z_0} \dfrac{1}{z^n} (z_0 \neq 0)$
   
   (b) $\lim\limits_{z \to i} \dfrac{iz^3 - 1}{z + i}$
   
   (c) $\lim\limits_{z \to z_0} \dfrac{P(z)}{Q(z)}$

8. Use mathematical induction and property of limits to show that

$$\lim_{z \to z_0} z^n = z_0^n$$

when $n$ is a positive integer ($n = 1, 2, \cdots$).

9. Show that the limit of the function

$$f(z) = \left(\frac{z}{\bar{z}}\right)^2$$

as $z$ tends to 0 does not exist.

10. Use definition of limit to prove that

if $\lim\limits_{z \to z_0} f(z) = \omega_0$, then $\lim\limits_{z \to z_0} |f(z)| = |\omega_0|$.

11. Write $\Delta z = z - z_0$ and show that

$$\lim_{z \to z_0} f(z) = \omega_0 \text{ if and only if } \lim_{\Delta z \to 0} f(z_0 + \Delta z) = \omega_0.$$

12. Show that

$$\lim_{z \to z_0} f(z) g(z) = 0 \text{ if } \lim_{z \to z_0} f(z) = 0$$

and if there exists a positive number $M$ such that $|g(z)| \leq M$ for all $z$ in some neighborhood of $z_0$.

# 4 The Extended Complex Plane and the Point at Infinity

## 4.1 The Spherical Representation, the Extended Complex Plane

For many purposes it is useful to extend the system **C** of complex number by introduction of a symbol $\infty$ represents infinity. In the plane there is no room for a point corresponding to $\infty$, but we can of course introduce an "ideal" point which we call the point at infinity. The complex plane together with this point is called the extended complex plane.

To visualize the point at infinity, one can think of a sphere is known as the Riemann sphere, and the correspondence is called a stereographic projection. By special convention, we have

(1) $a \pm \infty = \infty \pm a = \infty$;

(2) $b \cdot \infty = \infty \cdot b = \infty$ for all $b \neq 0$, including $b = \infty$;

(3) $\infty \pm \infty, 0 \cdot \infty, \dfrac{\infty}{\infty}, \dfrac{0}{0}$ is impossible;

(4) $\dfrac{\infty}{a} = \infty, \dfrac{a}{\infty} = 0$ for $a \neq \infty$;

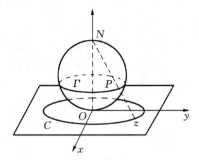

**Fig. 16**

(5) The real and imaginary part of $\infty$, the argument of $\infty$ is meaningless; $|\infty| = +\infty$;

(6) We agree that every straight line shall pass through the point at infinity. By contrast, no half plane shall contain the ideal point.

## 4.2 Some Concepts in the Extended Complex Plane

**Definition 1.4.1** For each small positive number $\varepsilon$, those points in the complex plane exterior to the circle $|z| = \dfrac{1}{\varepsilon}$ correspond to points on the sphere close to $N$. We call the set $D = \left\{z: |z| > \dfrac{1}{\varepsilon}\right\}$ an $\varepsilon$-neighborhood, or neighborhood, of $\infty$.

Let us agree that, in referring to a point $z$, we mean a point in the finite plane. Hereafter, when the point at infinity is to be considered, it will be specifically mentioned.

A meaning is now readily given to the statement
$$\lim_{z \to z_0} f(z) = w_0$$
when either $z_0$ or $w_0$, or possibly each of these numbers, is replaced by the point at infinity. In the definition of limit in Sec. 3, we simply replace the appropriate neighborhoods of $z_0$ and $w_0$ by neighborhoods of $\infty$. For example, $\lim_{z \to z_0} f(z) = \infty$ means that for each positive number $M$, there is a positive number $\delta$ such that
$$|f(z)| > M \text{ whenever } 0 < |z - z_0| < \delta.$$

And $\lim_{z\to\infty} f(z) = w_0$ means that for each positive number $\varepsilon$, there exists a positive number $M$ such that
$$|f(z) - w_0| < \varepsilon \text{ whenever } |z| > M.$$

**Theorem 1.4.1** If $z_0$ and $w_0$ are points in the $z$-plane and $w$-plane, respectively, then

$$\lim_{z\to z_0} f(z) = \infty \text{ if and only if } \lim_{z\to z_0} \frac{1}{f(z)} = 0 \qquad (1.4.1)$$

and

$$\lim_{z\to\infty} f(z) = w_0 \text{ if and only if } \lim_{z\to 0} f\left(\frac{1}{z}\right) = w_0 \qquad (1.4.2)$$

Moreover,

$$\lim_{z\to\infty} f(z) = \infty \text{ if and only if } \lim_{z\to 0} \frac{1}{f(1/z)} = 0 \qquad (1.4.3)$$

## Exercises

1. Use Theorem to show that

   (a) $\lim\limits_{z\to\infty} \dfrac{4z^2}{(z-1)^2} = 4$    (b) $\lim\limits_{z\to 1} \dfrac{1}{(z-1)^3} = \infty$    (c) $\lim\limits_{z\to\infty} \dfrac{z^2+1}{z-1} = \infty$

2. For each of the following points in **C**, give the corresponding point of $S$: $0$, $1+i$, $3+2i$.

3. Show that $z$ and $z'$ correspond to diametrically opposite points on the Riemann sphere if and only if $zz' = -1$.

4. State why limits involving the point at infinity are unique.

5. Show that a set $S$ is unbounded if and only if every neighborhood of the point at infinity contains at least one point in $S$.

# Chapter II

# Analytic Functions

The theory of functions of a complex variable aims at extending calculus to the complex domain. In this chapter, we introduce analytic functions, which play a central role in complex analysis.

## 1 The Concept of the Analytic Function

### 1.1 The Derivative of The Functions of a Complex Variable

**Definition 2.1.1** Let $f(z)$ be a function whose domain of definition contains a neighborhood of a point $z$. We say that $f(z)$ is differentiable at $z$ if the limit

$$f'(z) = \lim_{h \to 0} \frac{f(z+h) - f(z)}{h} \qquad (2.1.1)$$

exists. The value of this limit is denoted by $f'(z)$ or $\dfrac{\mathrm{d}f}{\mathrm{d}z}$ and is called the derivative of $f(z)$ at $z$.

In the complex variable case there are an infinity of directions in which a variable can approach a point $a$. In the real case, however, there are only two avenues of approach. Continuity, for example, of a function defined on **R** can be discussed in terms of right and left continuity; this is far from the case for functions of a complex variable. So the statement that a function of a complex variable has a derivative is stronger than the same statement about a function of a real variable.

The usual proofs in calculus concerning basic properties of differentiability are valid for complex differentiability.

**Theorem 2.1.1** If $f(z)$ is differentiable at $z$ then $f(z)$ is continuous at $z$.
**Proof** Because

$$\lim_{h\to 0}[f(z+h)-f(z)] = \lim_{h\to 0}\frac{f(z+h)-f(z)}{h}h$$

And since the limit of a product is the product of the limits, the limit on the right-hand side is equal to zero.

The converse is the continuity of a function at a point does not imply the existence of a derivative there.

**Theorem 2.1.2(Sum)** Let $f(z)$, $g(z)$ be a function whose domain of definition contains a neighborhood of a point $z$. We assume that $f(z)$, $g(z)$ are differentiable at $z$. Then the sum $f(z)+g(z)$ is differentiable at $z$, and

$$\frac{d}{dz}[f(z)+g(z)] = f'(z)+g'(z) \tag{2.1.2}$$

**Proof** This is immediate from the theorem that the limit of a sum is the sum of the limits.

**Theorem 2.1.3(Product)** Let $f(z)$, $g(z)$ be function whose domain of definition contains a neighborhood of a point $z$. We assume that $f(z)$, $g(z)$ are differentiable at $z$. Then the product $f(z)g(z)$ is differentiable at $z$, and

$$\frac{d}{dz}[f(z)g(z)] = f'(z)g(z)+f(z)g'(z) \tag{2.1.3}$$

**Proof** Because

$$\frac{d}{dz}[f(z)g(z)] = \lim_{h\to 0}\frac{f(z+h)g(z+h)-f(z)g(z)}{h}$$

$$= \lim_{h\to 0}\frac{f(z+h)g(z+h)-f(z)g(z+h)+f(z)g(z+h)-f(z)g(z)}{h}$$

$$= \lim_{h\to 0}\left[\frac{f(z+h)-f(z)}{h}g(z+h)+f(z)\frac{g(z+h)-g(z)}{h}\right]$$

By Definition 2.1.1 and the Theorem 2.1.1, we have

$$(fg)'(z) = \lim_{h\to 0}\frac{f(z+h)g(z+h)-f(z)g(z)}{h} = f'(z)g(z)+f(z)g'(z)$$

**Theorem 2.1.4(Quotient)** Let $f(z)$, $g(z)$ be a function whose domain of definition contains a neighborhood of a point $z$. We assume that $f(z)$, $g(z)$ are differentiable at $z$. Then the quotient $\frac{f(z)}{g(z)}$ is differentiable at $z$ provided $g(z)\neq 0$, and

$$\frac{d}{dz}\left[\frac{f(z)}{g(z)}\right] = \frac{f'(z)g(z)-f(z)g'(z)}{g^2(z)} \tag{2.1.4}$$

**Proof**  We first prove the differentiability of the quotient function $\dfrac{1}{g(z)}$. We have

$$\frac{d}{dz}\left[\frac{1}{g(z)}\right] = \lim_{h\to 0} \frac{\dfrac{1}{g(z+h)} - \dfrac{1}{g(z)}}{h}$$

$$= \lim_{h\to 0}\left[-\frac{g(z+h)-g(z)}{h} \cdot \frac{1}{g(z+h)g(z)}\right]$$

By Definition 2.1.1 and the Theorem 2.1.1, we have

$$\frac{d}{dz}\left[\frac{1}{g(z)}\right] = \lim_{h\to 0}\frac{\dfrac{1}{g(z+h)}-\dfrac{1}{g(z)}}{h} = -\frac{g'(z)}{g^2(z)}$$

The general formula for a quotient is obtained from this by writing

$$\frac{f(z)}{g(z)} = f(z)\frac{1}{g(z)}$$

And using the Theorem 2.1.3 and the derivative of $\dfrac{1}{g(z)}$.

**Theorem 2.1.5 (Chain Rule)**  Let $w=f(z)$. Assume that $f(z)$ is differentiable at $z$, and $g(z)$ is differentiable at $w$. Then $g[f(z)]$ is differentiable at $z$, and

$$\frac{d}{dz}\{g[f(z)]\} = g'[f(z)]f'(z) \qquad (2.1.5)$$

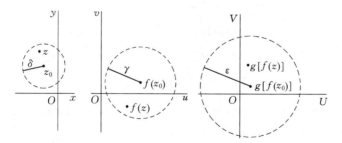

Fig. 17

**Proof**  The proof is the same as in calculus.

Let $w=f(z)$, and
$$k = f(z+h)-f(z)$$

So that

$$g[f(z+h)] - g[f(z)] = g(w+k) - g(w)$$

There exists a function $\psi(k)$ such that $\lim_{k \to 0} \psi(k) = 0$ and

$$g(w+k) - g(w) = g'(w)k + k\psi(k)$$
$$= g'(w)[f(z+h) - f(z)] + [f(z+h) - f(z)]\psi(k)$$

$$\lim_{h \to 0} \frac{g[f(z+h)] - g[f(z)]}{h} = \lim_{h \to 0} \frac{g(w+k) - g(w)}{h}$$

$$= g'(w) \frac{f(z+h) - f(z)}{h} + \frac{f(z+h) - f(z)}{h} \psi(k)$$

$$= g'[f(z)] f'(z)$$

**Example 1**  Suppose that $f(z) = z^n$. To show that $f'(z)$ exist everywhere.

At any point $z$,

$$\lim_{\Delta z \to 0} \frac{\Delta w}{\Delta z} = \lim_{\Delta z \to 0} \frac{(z+\Delta z)^n - z^n}{\Delta z} = n z^{n-1}$$

Hence $f'(z) = n z^{n-1}$.

In fact, the basic differentiation formulas can be derived from that definition by essentially the same steps as the ones used in calculus.

**Example 2**  Consider the function $f(z) = |z|^2$. To show that $f'(z)$ does not exist at any nonzero point (i. e., the function $f(z)$ has a derivative only at the origin).

Here

$$\frac{\Delta w}{\Delta z} = \frac{|z+\Delta z|^2 - |z|^2}{\Delta z} = \frac{(z+\Delta z)(\overline{z}+\overline{\Delta z}) - z\overline{z}}{\Delta z} = \overline{z} + \overline{\Delta z} + z \frac{\overline{\Delta z}}{\Delta z}$$

In particular, when $\Delta z$ approaches the origin horizontally through the points $(\Delta x, 0)$ on the real axis, we have

$$\lim_{\Delta z \to 0 \to 0} \frac{\Delta w}{\Delta z} = \lim_{\Delta z \to 0} (\overline{z} + \overline{\Delta z} + z) = \overline{z} + z$$

However, when $\Delta z$ approaches the origin vertically through the points $(0, \Delta y)$ on the imaginary axis, so that

$$\lim_{\Delta z \to 0} \frac{\Delta \omega}{\Delta z} = \lim_{\Delta z \to 0} (\overline{z} + \overline{\Delta z} - z) = \overline{z} - z$$

If the limit of $\frac{\Delta w}{\Delta z}$ exists, it may be found by letting the point $\Delta z = (\Delta x, \Delta y)$ approach the origin in the $\Delta z$ plane in any manner. So that

$$\overline{z} + z = \overline{z} - z$$

and so $z=0$, if $\dfrac{\mathrm{d}w}{\mathrm{d}z}$ exists.

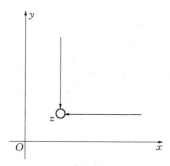

**Fig. 18**

Since the real and imaginary parts of $f(z)=|z|^2$ are
$$u(x,\ y)=x^2+y^2 \text{ and } v(x,\ y)=0$$
respectively, it shows that the real and imaginary components of a function of a complex variable can have continuous partial derivatives of all orders at a point and yet the function may not be differentiable there.

The function $f(z)=|z|^2$ is continuous at each point in plane since its components are continuous at each point.

**Definition 2.1.2** If $f(z)$ is differentiable at each point of a domain $D$ we say that $f(z)$ is differentiable on $D$. If $f'(z)$ is continuous then we say that $f(z)$ is continuously differentiable. If $f'(z)$ is differentiable then $f(z)$ is twice differentiable; continuing, a differentiable function such that each successive derivative is again differentiable is called infinitely differentiable.

## 1.2 Analytic Functions

**Definition 2.1.3** A function $f(z)$ of the complex variable $z$ is analytic in an open set if it has a derivative at each point in that set. If we should speak of a function $f(z)$ that is analytic in a set $S$ which is not open, it is to be understood that $f(z)$ is analytic in an open set containing $S$. In particular, $f(z)$ is analytic at a point $z_0$ if it is analytic throughout some neighborhood of $z_0$.

**Definition 2.1.4** An entire function is a function that is analytic at each point in the entire finite plane.

**Definition 2.1.5** If a function $f(z)$ fails to be analytic at a point $z_0$ but is analytic at some point in every neighborhood of $z_0$, then $z_0$ is called a singular point, or singularity, of $f(z)$.

**Theorem 2.1.6** If two functions are analytic on a domain $D$, their sum and their product are both analytic in $D$. And, their quotient is analytic on $D$ provided the function in the denominator does not vanish at any point in $D$.

**Theorem 2.1.7** A composition of two analytic functions is analytic.

**Theorem 2.1.8** If $f'(z)=0$ everywhere in a domain $D$, then $f(z)$ must be constant throughout $D$.

### Exercises

1. Suppose that $f(z_0)=g(z_0)=0$ and that $f'(z_0)$ and $g'(z_0)$ exist, where $g'(z_0)\neq 0$. Use definition of derivative to show that
$$\lim_{z\to z_0}\frac{f(z)}{g(z)}=\frac{f'(z_0)}{g'(z_0)}$$

2. To show that $f'(z)$ is not analytic at any point $z$ when
   (a) $f(z)=\bar{z}$     (b) $f(z)=\text{Re}z$     (c) $f(z)=\text{Im}z$
   (d) $f(z)=|z|$     (e) $f(z)=\dfrac{1}{\bar{z}}$     (f) $f(z)=xy^2+ix^2y$
   (g) $f(z)=x+y$     (h) $f(z)=x^2+iy^2$     (i) $f(z)=2x^3+3iy^3$

3. Prove that the functions $f(z)$ and $\overline{f(z)}$ are simultaneously analytic.

4. If $g(w)$ and $f(z)$ are analytic functions, show that $g(f(z))$ is also analytic.

## 2 Cauchy-Riemann Equations

**Theorem 2.2.1 (Necessary Conditions for Differentiability)** Suppose that
$$f(z)=u(x,y)+iv(x,y)$$
and that $f'(z_0)$ exists where $z_0=x_0+iy_0$. Then $u(x,y)$ and $v(x,y)$ must be differentiable at $(x_0, y_0)$, and they must satisfy the Cauchy-Riemann equations
$$u_x(x_0,y_0)=v_y(x_0,y_0), u_y(x_0,y_0)=-v_x(x_0,y_0) \qquad (2.2.1)$$
Also, $f'(z_0)$ can be written

$$f'(z_0) = u_x(x_0, y_0) + iv_x(x_0, y_0) \qquad (2.2.2)$$

**Proof** Suppose that
$$f(z) = u(x, y) + iv(x, y)$$
and that $f'(z_0)$ exists where $z_0 = x_0 + iy_0$.

We write $z_0 = x_0 + iy_0$, $\Delta z = \Delta x + i\Delta y$, $f'(z_0) = A + iB$ and
$$\Delta w = f(z_0 + \Delta z) - f(z_0)$$
$$= [u(x_0 + \Delta x, y_0 + \Delta y) - u(x_0, y_0)] +$$
$$i[v(x_0 + \Delta x, y_0 + \Delta y) - v(x_0, y_0)]$$

Assuming that the derivative
$$f'(z_0) = \lim_{\Delta z \to 0} \frac{\Delta w}{\Delta z} = \lim_{\Delta z \to 0} \text{Re} \frac{\Delta w}{\Delta z} + i \lim_{\Delta z \to 0} \text{Im} \frac{\Delta w}{\Delta z}$$
exists. We evaluate the limit in two different ways.

First let $(\Delta x, \Delta y)$ tend to $(0, 0)$ on the real axes in the $\Delta z$ plane, then
$$\lim_{\Delta z \to 0} \text{Re} \frac{\Delta w}{\Delta z} = \lim_{\Delta x \to 0} \frac{u(x_0 + \Delta x, y_0) - u(x_0, y_0)}{\Delta x} = u_x(x_0, y_0)$$
and
$$\lim_{\Delta z \to 0} \text{Im} \frac{\Delta w}{\Delta z} = \lim_{\Delta x \to 0} \frac{v(x_0 + \Delta x, y_0) - v(x_0, y_0)}{\Delta x} = v_x(x_0, y_0)$$
Thus
$$f'(z_0) = \lim_{\Delta z \to 0} \frac{\Delta w}{\Delta z} = u_x(x_0, y_0) + iv_x(x_0, y_0) \qquad (2.2.3)$$

However, we let $(\Delta x, \Delta y)$ tend to $(0, 0)$ on the imaginary axes in the $\Delta z$ plane, then
$$\lim_{\Delta z \to 0} \text{Re} \frac{\Delta w}{\Delta z} = \lim_{\Delta y \to 0} \frac{v(x_0, y_0 + \Delta y) - v(x_0, y_0)}{\Delta y} = v_y(x_0, y_0)$$
and
$$\lim_{\Delta z \to 0} \text{Im} \frac{\Delta w}{\Delta z} = -\lim_{\Delta y \to 0} \frac{u(x_0, y_0 + \Delta y) - u(x_0, y_0)}{\Delta y} = -u_y(x_0, y_0)$$
Thus
$$f'(z_0) = \lim_{\Delta z \to 0} \frac{\Delta w}{\Delta z} = v_y(x_0, y_0) - iu_y(x_0, y_0) \qquad (2.2.4)$$

Equating the real and imaginary parts of (2.2.3) and (2.2.4) we get the functions $u(x, y)$ and $v(x, y)$ are differentiable at the point $z_0 = (x_0, y_0)$ and

$$u_x(x_0, y_0) = v_y(x_0, y_0), \ u_y(x_0, y_0) = -v_x(x_0, y_0)$$

Furthermore, we have

$$f'(z_0) = u_x(x_0, y_0) + iv_x(x_0, y_0)$$

Equations (2.2.2) give $f'(z_0)$ in terms of partial derivatives of the component functions $u(x, y)$ and $v(x, y)$, but they also provide necessary conditions for the existence of $f'(z_0)$. Equations (2.2.1) are called the Cauchy-Riemann equations.

**Example 3** To show that the function

$$f(z) = z^2 = x^2 - y^2 + i2xy$$

is differentiable everywhere and that $f'(z) = 2z$. To verify that the Cauchy-Riemann equations are satisfied everywhere.

We note that

$$u(x, y) = x^2 - y^2 \text{ and } v(x, y) = 2xy$$

Thus

$$u_x = 2x = v_y, \ u_y = -2x = -v_x$$

Moreover, according to equation (2.2.2),

$$f'(z) = 2x + i2y = 2(x + iy) = 2z$$

Since the Cauchy-Riemann equations are necessary conditions for the existence of the derivative of a function $f(z)$ at a point $z_0$, they can often be used to locate points at which $f(z)$ does not have a derivative.

**Example 4** When $f(z) = |z|^2$, we have

$$u(x, y) = x^2 + y^2 \text{ and } v(x, y) = 0$$

If the Cauchy-Riemann equations are to hold at a point $(x, y)$, it follows that $2x = 0$ and $2y = 0$, or that $x = y = 0$. Consequently, $f'(z)$ does not exist at any nonzero point.

Satisfaction of the Cauchy-Riemann equations at a point $z_0 = (x_0, y_0)$ is not sufficient to ensure the existence of the derivative of a function $f(z)$ at that point. We have the following useful theorem.

**Theorem 2.2.2(Sufficient Conditions for Differentiability)** Let the function

$$f(z) = u(x, y) + iv(x, y)$$

be defined throughout some $\varepsilon$ neighborhood of a point $z_0 = x_0 + iy_0$, and suppose that the functions $u(x, y)$ and $v(x, y)$ are differentiable at $(x_0, y_0)$ and satisfy the Cauchy-Riemann equations

$$u_x = v_y, \quad u_y = -v_x$$

at $(x_0, y_0)$, then $f'(z_0)$ exists.

**Proof** We write $\Delta z = \Delta x + i\Delta y$, where $0 < |\Delta z| < \varepsilon$, and
$$\Delta w = f(z_0 + \Delta z) - f(z_0)$$

Thus
$$\Delta w = \Delta u + i\Delta v \tag{2.2.5}$$

where
$$\Delta u = u(x_0 + \Delta x, y_0 + \Delta y) - u(x_0, y_0)$$

and
$$\Delta v = v(x_0 + \Delta x, y_0 + \Delta y) - v(x_0, y_0)$$

The assumption that $u(x, y)$ and $v(x, y)$ are differentiable at the point $(x_0, y_0)$ enables us to write
$$\Delta u = u_x(x_0, y_0)\Delta x + u_y(x_0, y_0)\Delta y + \alpha \tag{2.2.6}$$

and
$$\Delta v = v_x(x_0, y_0)\Delta x + v_y(x_0, y_0)\Delta y + \beta \tag{2.2.7}$$

where $\alpha$ and $\beta$ tend to 0 as $(\Delta x, \Delta y)$ approaches $(0, 0)$. By expressions (2.2.5), (2.2.6), (2.2.7), we have
$$\Delta w = u_x(x_0, y_0)\Delta x + u_y(x_0, y_0)\Delta y + \alpha + i[v_x(x_0, y_0)\Delta x + v_y(x_0, y_0)\Delta y + \beta] \tag{2.2.8}$$

Assuming that the Cauchy-Riemann equations are satisfied at $(x_0, y_0)$, we can get
$$\frac{\Delta w}{\Delta z} = u_x(x_0, y_0) + iv_x(x_0, y_0) + \frac{\alpha + i\beta}{\Delta z} \tag{2.2.9}$$

This means that the limit of the left-hand side of equation (2.2.9) exists and that
$$f'(z_0) = u_x(x_0, y_0) + iv_x(x_0, y_0) \tag{2.2.10}$$

Since a real-valued function of two real variables with continuous partial derivatives is differentiable, we get from Theorem 2.2.2 that

**Corollary 2.2.3** Let the function
$$f(z) = u(x, y) + iv(x, y)$$
be defined on a domain $D$ and suppose that the functions $u(x, y)$ and $v(x, y)$ are differentiable and satisfy the Cauchy-Riemann equations at every point of $D$, then $f(z)$ is differentiable on $D$ and

$$f'(z) = u_x(x, y) + iv_x(x, y), \quad \forall z = x + iy \in D$$

A combination of Theorem 1 and Theorem 2 gives a necessary and sufficient condition for a function to be differentiable, which is given by the following corollary.

**Corollary 2.2.4** A function $f(z) = u(x, y) + iv(x, y)$ is differentiable at a point $z_0 = x_0 + iy_0$ if and only if the functions $u(x, y)$ and $v(x, y)$ are differentiable at $(x_0, y_0)$ and satisfy the Cauchy-Riemann equations

$$u_x(x_0, y_0) = v_y(x_0, y_0), \quad u_y(x_0, y_0) = -v_x(x_0, y_0)$$

**Theorem 2.2.5** If $u(x, y)$ and $v(x, y)$ have continuous first-order partial derivatives which satisfy the Cauchy-Riemann equations

$$u_x(x, y) = v_y(x, y), \quad u_y(x, y) = -v_x(x, y)$$

Then

$$f(z) = u(x, y) + iv(x, y)$$

is analytic with continuous derivative $f'(z)$, and conversely.

**Example 5** It also follows from Corollary 2.2.3 that the function $f(z) = |z|^2$, whose components are

$$u(x, y) = x^2 + y^2 \text{ and } v(x, y) = 0$$

has a derivative at $z = 0$. $f'(0) = 0 + i0 = 0$.

We saw that this function cannot have a derivative at any nonzero point since the Cauchy-Riemann equations are not satisfied at such points.

**Example 6** Suppose that a function

$$f(z) = u(x, y) + iv(x, y)$$

and its conjugate

$$\overline{f(z)} = u(x, y) - iv(x, y)$$

are both analytic on a given domain $D$. To show that $f(z)$ must be constant throughout $D$.

**Proof** Because of the analyticity of $f(z)$, the Cauchy-Riemann equations

$$u_x = v_y, \quad u_y = -v_x \tag{2.2.11}$$

hold in $D$, according to the theorem. Also, the analyticity of $\overline{f(z)}$ in $D$ tells us that

$$u_x = (-v)_y = -v_y, \quad u_y = -(-v_x) = v_x \tag{2.2.12}$$

By adding corresponding sides of the first of equations (2.2.11) and (2.2.12), we find that $u_x = 0$ and $v_x = 0$ in $D$. According to expression

(2.2.10), then,
$$f'(z)=u_x+iv_x=0+i0=0$$
and it follows from the theorem that $f(z)$ is constant throughout $D$.

**Theorem 2.2.6** An analytic function in a region $D$ whose derivative vanishes identically must reduce to a constant. The same is true if either the real part, the imaginary part, the modulus, or the argument is constant.

**Proof**

(1) The vanishing of the derivative implies that $\frac{\partial u}{\partial x}$, $\frac{\partial u}{\partial y}$, $\frac{\partial v}{\partial x}$, $\frac{\partial v}{\partial y}$ are all zero. It follows that $u$ and $v$ are constant. We conclude that $u+iv$ is constant.

(2) If $u$ or $v$ is constant,
$$f'(z)=\frac{\partial u}{\partial x}-i\frac{\partial u}{\partial y}=\frac{\partial v}{\partial y}+i\frac{\partial v}{\partial x}=0$$
And hence $f(z)$ must be constant.

(3) If $u^2+v^2$ is constant, we obtain
$$u\frac{\partial u}{\partial x}+v\frac{\partial v}{\partial x}=0$$
and
$$u\frac{\partial u}{\partial y}+v\frac{\partial v}{\partial y}=-u\frac{\partial v}{\partial x}+v\frac{\partial u}{\partial x}=0$$
These equations permit the conclusion
$$\frac{\partial u}{\partial x}=\frac{\partial v}{\partial x}=0$$
Unless the determinant $u^2+v^2$ vanishes. But if $u^2+v^2=0$ at a single point it is constantly zero and $f(z)$ vanishes identically. Hence $f(z)$ is any case a constant.

(4) Finally, if $\arg f(z)$ is constant, we can set $u=kv$ with constant $k$ (unless $v$ is identically zero). But $u-kv$ is the real part of $(1+ik)f(z)$, and we conclude again that $f(z)$ must reduce to a constant.

Note that for this theorem it is essential that $D$ is a region. If not, we can only assert that $f(z)$ is constant on each component of $D$.

## Exercises

1. Verify Cauchy-Riemann's equations for functions $z^2$, $z^3$.

2. To show that $f'(z)$ and its derivative $f''(z)$ exist everywhere, and find $f''(z)$ when
   (a) $f(z) = iz + 2$
   (b) $f(z) = e^{-x}e^{-iy}$
   (c) $f(z) = z^3$
   (d) $f(z) = \cos x \mathrm{chy} - i\sin x \mathrm{shy}$

3. To determine where $f'(z)$ exists and find its value when
   (a) $f(z) = 1/z$
   (b) $f(z) = x^2 + iy^2$
   (c) $f(z) = z\mathrm{Im}z$

4. To verify that each of these functions is entire and find its derivative $f'(z)$
   (a) $f(z) = x^3 + 3x^2yi - 3xy^2 - y^3 i$
   (b) $f(z) = e^x(x\cos y - y\sin y) + ie^x(y\cos y + x\sin y)$
   (c) $f(z) = \sin x \mathrm{chy} + i\cos x \mathrm{shy}$
   (d) $f(z) = \cos x \mathrm{chy} - i\sin x \mathrm{shy}$

5. Show that an analytic functions cannot have a constant absolute value without reducing to a constant.

6. Let a function $f(z)$ be analytic in a domain $D$. Prove that $f(z)$ must be constant throughout $D$ if one of the following holds:
   (a) $f(z)$ is real-valued for all $z$ in $D$
   (b) $|f(z)|$ is constant throughout $D$

7. Suppose that $f(z)$ is analytic and satisfies the condition $|f^2(z)-1|<1$ in a region $D$. Show that either $\mathrm{Re}f(z)>0$ or $\mathrm{Re}f(z)<0$ thought out $D$.

8. Let the function $f(z)$ be analytic in a domain $D$. Prove that
   (a) $\left(\dfrac{\partial}{\partial x}|f(z)|\right)^2 + \left(\dfrac{\partial}{\partial y}|f(z)|\right)^2 = |f'(z)|^2$
   (b) $\left(\dfrac{\partial^2}{\partial x^2} + \dfrac{\partial^2}{\partial y^2}\right)|f(z)|^2 = 4|f'(z)|^2$

# 3　Elementary Functions

　　In this section, we consider various elementary functions studied in calculus and define corresponding functions of a complex variable. To be specific, we define analytic functions of a complex variable $z$ that reduce to the elementary functions in calculus when $z = x + i0$.

## 3.1　The Exponential Function

**Definition 2.3.1**　We define the exponential function $e^z$ by writing

Chapter II  Analytic Functions

$$e^z = e^x e^{iy}, \quad \forall z = x+iy \in \mathbf{C} \tag{2.3.1}$$

We see from this definition that $e^z$ reduces to the usual exponential function in calculus when $y=0$; and, following the convention used in calculus, we often write $\exp z$ for $e^z$.

There are a number of other important properties of $e^z$ that are expected:

(1) $$e^{z_1} e^{z_2} = e^{z_1+z_2} \tag{2.3.2}$$

(2) $$\frac{e^{z_1}}{e^{z_2}} = e^{z_1-z_2} \tag{2.3.3}$$

(3) $$\frac{1}{e^z} = e^{-z} \tag{2.3.4}$$

(4) $$\frac{d}{dz} e^z = e^z \tag{2.3.5}$$

(5) The differentiability of $e^z$ for all $z$ tells us that $e^z$ is entire. It is also true that

$$e^z \neq 0 \text{ for any complex number } z \tag{2.3.6}$$

(6) $\quad |e^z| = e^x$ and $\text{Arg}(e^z) = y + 2n\pi (n = 0, \pm 1, \pm 2, \cdots)$ $\quad$ (2.3.7)

**Definition 2.3.2** A function $f(z)$ is said to be periodic with period $\omega \neq 0$ if
$$f(z+\omega) = f(z)$$

(7) $e^z$ is periodic, with a pure imaginary period $2\pi i$:

$$e^{z+2\pi i} = e^z, \forall z \in \mathbf{C} \tag{2.3.8}$$

**Example 7** To find the values of $z$ such that

$$e^z = -1 \tag{2.3.9}$$

We write equation (2.3.9) as $e^x e^{iy} = 1 e^{i\pi}$. Then, in view of the statement that regarding the equality of two nonzero complex numbers in exponential form,

$$e^x = 1 \quad \text{and} \quad y = \pi + 2n\pi (n=0, \pm 1, \pm 2, \cdots)$$

Thus, $x = 0$, and we find that

$$z = (2n+1)\pi i \quad (n = 0, \pm 1, \pm 2, \cdots)$$

## 3.2  Trigonometric Functions

**Definition 2.3.3** To define the sine and cosine functions of a complex variable $z$ as follows:

$$\sin z = \frac{e^{iz} - e^{-iz}}{2i}, \cos z = \frac{e^{iz} + e^{-iz}}{2} \tag{2.3.10}$$

These functions are entire since they are linear combinations of the entire functions $e^{iz}$ and $e^{-iz}$. A number of important properties of $\sin z$ and $\cos z$ follow immediately:

(1) $$\frac{d}{dz}\sin z = \cos z, \quad \frac{d}{dz}\cos z = -\sin z \qquad (2.3.11)$$

(2) $$\sin(-z) = -\sin z \text{ and } \cos(-z) = \cos z \qquad (2.3.12)$$

(3) $$2\sin z_1 \cos z_2 = \sin(z_1+z_2) + \sin(z_1-z_2) \qquad (2.3.13)$$
$$\sin(z_1+z_2) = \sin z_1 \cos z_2 + \cos z_1 \sin z_2 \qquad (2.3.14)$$
$$\cos(z_1+z_2) = \cos z_1 \cos z_2 - \sin z_1 \sin z_2 \qquad (2.3.15)$$
$$\sin 2z = 2\sin z \cos z, \quad \cos 2z = \cos^2 z - \sin^2 z \qquad (2.3.16)$$
$$\sin\left(z+\frac{\pi}{2}\right) = \cos z, \quad \sin\left(z-\frac{\pi}{2}\right) = -\cos z \qquad (2.3.17)$$

(4) The periodic character of these functions is evident:
$$\sin(z+2\pi) = \sin z, \quad \sin(z+\pi) = -\sin z,$$
$$\cos(z+\pi) = \cos z, \quad \cos(z+2\pi) = \cos z \qquad (2.3.18)$$

(5) When $y$ is any real number, the hyperbolic functions
$$\sh y = \frac{e^y - e^{-y}}{2} \text{ and } \ch y = \frac{e^y + e^{-y}}{2} \qquad (2.3.19)$$

from calculus, we write
$$\sin(iy) = i\sh y \text{ and } \cos(y) = \ch y \qquad (2.3.20)$$
$$\sin z = \sin x \ch y + i\cos x \sh y \qquad (2.3.21)$$
$$\cos z = \cos x \ch y - i\sin x \sh y \qquad (2.3.22)$$
$$|\sin z|^2 = \sin^2 x + \sh^2 y, \quad |\cos z|^2 = \cos^2 x + \sh^2 y \qquad (2.3.23)$$

where $z = x + iy$.

(6) In as much as $\sh y$ tends to infinity as $y$ tends to infinity, it is clear from these two equations to infinity, it is clear from these two equations $\sin z$ and $\cos z$ are not bounded on the complex plane, whereas the absolute values of $\sin z$ and $\cos z$ are less than or equal to unity for all values of $z = x$.

(7) A zero of a given function $f(z)$ is a number $z_0$ such that $f(z_0) = 0$. Since $\sin z$ becomes the usual sine function in calculus when $z$ is real, we know that the real numbers $z = n\pi (n = 0, \pm 1, \pm 2, \cdots)$ are all zeros of $\sin z$. Since

$$\cos z = -\sin\left(z - \frac{\pi}{2}\right)$$

## Chapter II  Analytic Functions

we see that

$$\cos z = 0 \quad \text{if and only if} \quad z = \frac{\pi}{2} + n\pi (n=0, \pm 1, \pm 2, \cdots) \tag{2.3.24}$$

So, as was the case with $\sin z$, the zeros, of $\cos z$ are all real.

**Definition 2.3.4**  The other four trigonometric functions are defined in terms of the sine and cosine functions by the usual relations:

$$\tan z = \frac{\sin z}{\cos z}, \quad \cot z = \frac{\cos z}{\sin z} \tag{2.3.25}$$

$$\sec z = \frac{1}{\cos z}, \quad \csc z = \frac{1}{\sin z} \tag{2.3.26}$$

We list some properties:

(1) 
$$\frac{d}{dz}\tan z = \sec^2 z, \quad \frac{d}{dz}\cot z = -\csc^2 z \tag{2.3.27}$$

$$\frac{d}{dz}\sec z = \sec z \tan z, \quad \frac{d}{dz}\csc z = -\csc z \cot z \tag{2.3.28}$$

(2)  $\tan(z+\pi) = \tan z, \quad \forall z \in \mathbb{C}$ \hfill (2.3.29)

Thus, tangent function is $\pi$-periodic.

### 3.3  Hyperbolic Functions

**Definition 2.3.5**  The hyperbolic sine and the hyperbolic cosine of a complex variable are defined as they are with a real variable; that is,

$$\mathrm{sh}\, z = \frac{e^z - e^{-z}}{2}, \quad \mathrm{ch}\, z = \frac{e^z + e^{-z}}{2} \tag{2.3.30}$$

Since $e^z$ and $e^{-z}$ are entire, then $\mathrm{sh}\, z$ and $\mathrm{ch}\, z$ are entire. Furthermore, some of the most frequently used identities involving hyperbolic sine and cosine functions are

$$\frac{d}{dz}\mathrm{sh}\, z = \mathrm{ch}\, z, \quad \frac{d}{dz}\mathrm{ch}\, z = \mathrm{sh}\, z \tag{2.3.31}$$

$$-i\,\mathrm{sh}(iz) = \sin z, \quad \mathrm{ch}(iz) = \cos z \tag{2.3.32}$$

$$-i\sin(iz) = \mathrm{sh}\, z, \quad \cos(iz) = \mathrm{ch}\, z \tag{2.3.33}$$

$$\mathrm{ch}(-z) = \mathrm{ch}\, z \tag{2.3.34}$$

$$\mathrm{ch}^2 z - \mathrm{sh}^2 z = 1 \tag{2.3.35}$$

$$\mathrm{sh}(z_1 + z_2) = \mathrm{sh}\, z_1 \mathrm{ch}\, z_2 + \mathrm{ch}\, z_1 \mathrm{sh}\, z_2 \tag{2.3.36}$$

$$\mathrm{ch}(z_1 + z_2) = \mathrm{ch}\, z_1 \mathrm{ch}\, z_2 + \mathrm{sh}\, z_1 \mathrm{sh}\, z_2 \tag{2.3.37}$$

$$\text{sh}z = \text{sh}x\cos y + i\text{ch}x\sin y, \quad \text{ch}z = \text{ch}x\cos y + i\text{sh}x\sin y \quad (2.3.38)$$
$$|\text{sh}z|^2 = \text{sh}^2 x + \sin^2 y, \quad |\text{ch}z|^2 = \text{sh}^2 x + \cos^2 y \quad (2.3.39)$$

where $z = x + iy$.

**Definition 2.3.6** The hyperbolic tangent and sech$z$ of $z$ is defined by the equation

$$\text{th}z = \frac{\text{sh}z}{\text{ch}z}, \quad \text{sech}z = \frac{1}{\text{ch}z} \quad (2.3.40)$$

and is analytic in every domain in which $\text{ch}z \neq 0$. The hyperbolic cth$z$ and csch$z$ of $z$ is defined by the equation

$$\text{cth}z = \frac{\text{ch}z}{\text{sh}z} = \frac{1}{\text{th}z}, \quad \text{csch}z = \frac{1}{\text{sh}z} \quad (2.3.41)$$

and is analytic in every domain in which $\text{sh}z \neq 0$.

It is straightforward to verify the following differentiation formulas, which are the same as those established in calculus for the corresponding functions of a real variable:

$$\frac{d}{dz}\text{th}z = \text{sech}^2 z, \quad \frac{d}{dz}\text{cth}z = -\text{csch}^2 z,$$
$$\frac{d}{dz}\text{sech}z = -\text{sech}z\,\text{th}z, \quad \frac{d}{dz}\text{csch}z = -\text{csch}z\,\text{cth}z \quad (2.3.42)$$

### Exercises

1. Find the solutions of each of the following equations:
   (a) $e^z = -2$      (b) $e^z = 1 + \sqrt{3}i$      (c) $e^{2z-1} = 1$

2. To show that, for all $z$,
   (a) $\overline{\sin z} = \sin \bar{z}$      (b) $\overline{\cos z} = \cos \bar{z}$

3. To show that
   (a) $\overline{\cos(iz)} = \cos(i\bar{z})$    for all $z$
   (b) $\overline{\sin(iz)} = \sin(i\bar{z})$    if and only if $z = n\pi i (n = 0, \pm 1, \pm 2, \cdots)$

4. Show that
   (a) $e^{2 \pm 3\pi i} = -e^2$      (b) $e^{\frac{2+\pi i}{4}} = \sqrt{\frac{e}{2}}(1+i)$      (c) $e^{z+\pi i} = -e^z$

5. State why the function $2z^2 - 3 - ze^z + e^{-z}$ is entire.

6. Use the Cauchy-Riemann equations to show that the function $f(z) = \overline{e^z}$ is not analytic anywhere.

## Chapter II  Analytic Functions

7. Show in two ways that the function $e^{z^2}$ is entire. What is its derivative?
8. Write $|e^{2z+i}|$ and $|e^{iz^2}|$ in terms of $x$ and $y$. Then show that
$$|e^{2z+i}+e^{iz^2}| \leqslant e^{2x}+e^{-2xy}$$
9. Show that $|e^{z^2}| \leqslant e^{|z|^2}$.
10. Prove that $|e^{-2z}| \leqslant 1$ if and only if $\text{Re} z \geqslant 0$.
11. Describe the following sets:
    (a) $\{z: e^z=-1\}$  (b) $\{z: e^z=-i\}$  (c) $\{z: \cos z=0\}$.
12. Show that $\overline{e^{iz}}=e^{i\bar{z}}$ if and only if $z=n\pi (n=0, \pm 1, \pm 2, \cdots)$.
13. (a) Show that if $e^z$ is real, then $\text{Im} z=n\pi (n=0, \pm 1, \pm 2, \cdots)$;
    (b) If $e^z$ is pure imaginary, what restriction is placed on $z$?
14. Describe the behavior of $e^z=e^x e^{iy}$ as
    (a) $x$ tends to $-\infty$    (b) $y$ tends to $\infty$
15. Write $\text{Re}(e^{\frac{1}{z}})$ in terms of $x$ and $y$. Why is this function harmonic in every domain that does not contain the origin?
16. Find the real and imaginary parts of $e^{e^z}$.
17. To show that
    (a) $1+\tan^2 z=\sec^2 z$        (b) $1+\cot^2 z=\csc^2 z$
    (c) $|\text{sh } y| \leqslant |\sin z| \leqslant \text{ch } y$    (d) $|\text{sh } y| \leqslant |\cos z| \leqslant \text{ch } y$
18. Use the Cauchy-Riemann equations to show that neither $\overline{\sin z}$ nor $\overline{\cos z}$ is an analytic function of $z$ anywhere.
19. To show that, for all $z$,
    (a) $\overline{\sin z}=\sin \bar{z}$    (b) $\overline{\cos z}=\cos \bar{z}$
20. Show that
    (a) $\overline{\cos(iz)}=\cos(i\bar{z})$ for all $z$
    (b) $\overline{\sin(iz)}=\sin(i\bar{z})$ if and only if $z=n\pi i(n=0, \pm 1, \pm 2, \cdots)$
21. Find all roots of the equation $\sin z=\text{ch} 4$ by equating the real parts and the imaginary parts of $\sin z$ and $\text{ch} 4$.
22. Find all roots of the equation $\cos z=2$.
23. Show that
    (a) $\text{sh}(z+\pi i)=-\text{sh } z$
    (b) $\text{ch}(z+\pi i)=-\text{ch } z$
    (c) $\text{th}(z+\pi i)=\text{th } z$

# 4  Multi-Valued Functions

## 4.1  The Logarithmic Function

(1) Definition

**Definition 2.4.1**  We define the logarithmic function is the inverse function of the exponential function.

By definition, $w = \log z$ is a root of the equation $e^w = z$.

When $z$ and $w$ are written $z = re^{i\theta}\ (-\pi < \theta \leq \pi)$ and $w = u + iv$, then $e^u e^{iv} = re^{i\theta}$. Thus regarding the equality of two complex numbers expressed in exponential form.

$$e^u = r \quad \text{and} \quad v = \theta + 2n\pi$$

where $n$ is any integer. Since the equation $e^u = r$ is the same as $u = \ln r$, it follows that

$$w = \ln r + i(\theta + 2n\pi) \quad (n = 0, \pm 1, \pm 2, \cdots)$$

Thus, if we write

$$\text{Log} z = \ln r + i(\theta + 2n\pi) \quad (n = 0, \pm 1, \pm 2, \cdots) \qquad (2.4.1)$$

we have the simple relation

$$e^{\text{Log} z} = z \quad (z \neq 0) \qquad (2.4.2)$$

Thus, we get a multi-valued function. Sometimes, we write

$$\text{Log} z = \{\ln |z| + i(\arg z + 2n\pi): n \in \mathbb{Z}\}, \quad z \in \mathbb{C} \setminus \{0\},$$

which is indeed a set-valued function.

**Example 8**  If $z = -1 - \sqrt{3}i$, then $r = 2$ and $\theta = -\dfrac{2\pi}{3}$. Hence

$$\text{Log}(-1-\sqrt{3}i) = \ln 2 + i\left(-\frac{2\pi}{3} + 2n\pi\right) = \ln 2 + 2\left(n - \frac{1}{3}\right)\pi i \quad (n = 0, \pm 1, \pm 2, \cdots)$$

**Example 9**  From expression (2.4.1), we find that

$$\text{Log} 1 = \ln 1 + i(0 + 2n\pi) = 2n\pi i \quad (n = 0, \pm 1, \pm 2, \cdots)$$

and $\log 1 = 0$.

**Example 10**  Observe that

$$\text{Log}(-1) = \ln 1 + i(\pi + 2n\pi) = (2n+1)\pi i \quad (n = 0, \pm 1, \pm 2, \cdots)$$

and that $\log(-1) = \pi i$.

**Example 11**  Since

$$1-i=\sqrt{2}\,e^{-\frac{\pi}{4}i}$$

so

$$\log(1-i)=\frac{1}{2}\ln 2-\frac{\pi}{4}i+2n\pi i$$

The examples here reminds us that, although we were unable to find logarithms of negative real numbers in calculus, we can now do so.

(2) Branches and Derivatives of Logarithms(see reference[2])

If $z=re^{i\theta}$ is a nonzero complex number, the argument $\theta$ has any one of the values $\theta=\Theta+2n\pi$ ($n=0, \pm 1, \pm 2, \cdots$), where $\Theta=\arg z$. Hence the definition

$$\text{Log}\,z=\ln r+i(\Theta+2n\pi) \quad (n=0, \pm 1, \pm 2, \cdots)$$

of the multiple-valued logarithmic function can be written

$$\text{Log}\,z=\ln r+i\theta=\ln|z|+i\text{Arg}\,z \qquad (2.4.3)$$

If we let $\alpha$ denote any real number and restrict the value of $\theta$ in expression (2.4.3) so that $\alpha<\theta<\alpha+2\pi$, then we obtain a function

$$L_\alpha(z)=\ln r+i\theta \quad (z=re^{i\theta},\ r>0,\ \alpha<\theta<\alpha+2\pi) \qquad (2.4.4)$$

thus,

$$\frac{d}{dz}L_\alpha(z)=\frac{1}{z} \quad (|z|>0,\ \alpha<\text{Arg}\,z<\alpha+2\pi) \qquad (2.4.5)$$

In particular,

$$\frac{d}{dz}L_{-\pi}(z)=\frac{1}{z} \quad (|z|>0,\ -\pi<\arg z<\pi) \qquad (2.4.6)$$

For each fixed $\alpha$, the single-valued function (2.4.4) is a branch of the multiple-valued function (2.4.3). Especially, the branch

$$\log z=\ln|z|+i\arg z\ (z\neq 0) \qquad (2.4.7)$$

of the logarithmic function Log $z$ is called the principal branch.

(3) Some Identities Involving Logarithms

Some identities involving logarithms in calculus carry over to complex analysis and others do not.

If $z_1$ and $z_2$ denote any two nonzero complex numbers, it is straightforward to show that

$$\text{Log}(z_1 z_2)=\text{Log}\,z_1+\text{Log}\,z_2 \qquad (2.4.8)$$

This statement, involving a multiple-valued function, is to be interpreted in

the same way that the statement
$$\text{Arg}(z_1 z_2) = \text{Arg} z_1 + \text{Arg} z_2 \quad (2.4.9)$$
That is, if values of two of the three logarithms are specified, then there is value of the third logarithm such that equation (2.4.8) holds.

**Example 12** To illustrate statement (2.4.8), write $z_1 = z_2 = -1$ and note that $z_1 z_2 = 1$. If the values $\text{Log} z_1 = \pi i$ and $\text{Log} z_2 = -\pi i$ are specified, equation (2.4.8) is evidently satisfied when the value $\text{Log}(z_1 z_2) = 0$ is chosen.

Observe that, for the same numbers $z_1$ and $z_2$,
$$\log(z_1 z_2) = 0 \text{ and } \log z_1 + \log z_2 = 2\pi i$$
Thus statement (2.4.8) is not, in general, valid when Log is replaced everywhere by log.

Verification of the statement
$$\text{Log}\frac{z_1}{z_2} = \text{Log} z_1 - \text{Log} z_2 \quad (2.4.10)$$
which is to be interpreted in the same way as statement (2.4.8), is left to the exercises.

We include here two other properties of $\text{Log} z$ that will be of special interest in
$$z^n = e^{n \text{Log} z} \quad (n=0, \pm 1, \pm 2, \cdots) \quad (2.4.11)$$
for any value of $\text{Log} z$ that is taken. When $n=1$, this reduces, of course, Equation (2.4.11) is readily verified by writing $z = re^{i\theta}$ and noting that each side becomes $r^n e^{in\theta}$. Also,
$$z^{\frac{1}{n}} = e^{\frac{1}{n} \text{Log} z} \quad (n=1, 2, \cdots) \quad (2.4.12)$$

## 4.2 Complex Power Functions

**Definition 2.4.2** When $c$ is any complex number, the complex power function $z^c$ is defined by means of the equation
$$z^c = e^{c \text{Log} z}, \quad z \neq 0 \quad (2.4.13)$$
Equation (2.4.13) provides a consistent definition of $z^c$ in the sense that it is already known to be valid when $c = n$ ($n=0, \pm 1, \pm 2, \cdots$) and $c = 1/n$ ($n = \pm 1, \pm 2, \cdots$). Definition (2.4.2) is, in fact, suggested by those particular choices of $c$.

**Example 13** Powers of $z$ are, in general, multiple-valued, as illustrated

by writing
$$i^{-2i} = e^{-2i\text{Log}i}$$
and then
$$\text{Log}i = \ln 1 + i\left(\frac{\pi}{2} + 2n\pi\right) = i\left(2n + \frac{1}{2}\right)\pi \quad (n = 1, \pm 1, \pm 2, \cdots)$$
This shows that
$$i^{-2i} = e^{(4n+1)\pi} \quad (n = 0, \pm 1, \pm 2, \cdots)$$

Since the exponential function has the property $\dfrac{1}{e^z} = e^{-z}$, one can see that

$$\frac{1}{z^c} = z^{-c} \tag{2.4.14}$$

$$\frac{d}{dz}z^c = cz^{c-1} \quad (|z| > 0, \ \alpha < \text{Arg}z < \alpha + 2\pi) \tag{2.4.15}$$

The principal value of $z^c$ occurs when $\text{Log}z$ is replaced by $\log z$ in definition (2.4.13)

$$\text{P. V. } z^c = e^{c\log z} \tag{2.4.16}$$

Equation (2.4.16) also serves to define the principal branch of the function $z^c$ on the domain

$$D_{-\pi} = \{z : \ |z| > 0, \ -\pi < \arg z < \pi\}$$

**Example 14** The principal value of $\log i = \dfrac{\pi}{2}i$, $\log(-i) = -\dfrac{1}{2}\pi i$.

**Example 15** The principal value of $(-i)^i$ is
$$e^{i\log(-i)} = e^{i(\ln 1 - i\frac{\pi}{2})} = e^{\frac{\pi}{2}}$$
That is,
$$\text{P. V. }(-i)^i = e^{\frac{\pi}{2}}$$

**Example 16** The principal branch of $z^{\frac{2}{3}}$ can be written
$$e^{\frac{2}{3}\log z} = e^{\frac{2}{3}\ln r + i\frac{2}{3}\arg z} = \sqrt[3]{r^2}\, e^{i\frac{2}{3}\arg z}$$
Thus
$$\text{P. V. } z^{\frac{2}{3}} = \sqrt[3]{r^2}\cos\frac{2\arg z}{3} + i\sqrt[3]{r^2}\sin\frac{2\arg z}{3}$$

**Definition 2.4.3** According to definition (2.4.1), the exponential function with base $c$, where $c$ is any nonzero complex constant, is written

## Functions of Complex Variables

$$c^z = e^{z \text{Log} c} \tag{2.4.17}$$

Note that although $e^z$ is, in general, multiple-valued according to definition (2.4.17), the usual interpretation of $e^z$ occurs when the principal value of the logarithm is taken. For the principal value of $\log e$ is unity.

When a value of $\text{Log} c$ is specified, $e^z$ is an entire function of $z$. In fact,

$$\frac{d}{dz} c^z = \frac{d}{dz} e^{z \text{Log} c} = e^{z \text{Log} c} \text{Log} c$$

and this shows that

$$\frac{d}{dz} c^z = c^z \text{Log} c \tag{2.4.18}$$

### 4.3 Inverse Trigonometric and Hyperbolic Functions

**Definition 2.4.4** We define the inverse trigonometric function is the inverse function of the trigonometric function.

In order to define the inverse cosine function Arccos, we write

$$w = \text{Arccos} z$$

When

$$z = \frac{e^{iw} + e^{-iw}}{2}$$

If we put this equation in the form

$$(e^{iw})^2 - 2z(e^{iw}) + 1 = 0$$

We find that

$$e^{iw} = z \pm \sqrt{z^2 - 1} \tag{2.4.19}$$

where $\sqrt{z^2 - 1}$ is a double-valued function of $z$. Taking logarithms of each side of equation (2.4.19), we derive the expression

$$\text{Arccos} z = -i \text{Log}[z \pm \sqrt{z^2 - 1}] \tag{2.4.20}$$

Similarly, we can derive expression for Arcsin$z$ and Arctan$z$ to

$$\text{Arcsin} z = -i \text{Log}[iz \pm \sqrt{1 - z^2}] \tag{2.4.21}$$

and that

$$\text{Arctan} z = \frac{1}{2i} \text{Log} \frac{1 + iz}{1 - iz} \tag{2.4.22}$$

The functions Arccos, Arcsin$z$ and Arctan$z$ are multiple-values.

**Example 17** To compute $\text{Arcsin}(-i)$.

Expression (2.4.21) tells us that
$$\text{Arcsin}(-i)=-\text{Log}(1\pm\sqrt{2})$$
But
$$\text{Log}(1+\sqrt{2})=\ln(1+\sqrt{2})+2n\pi i \quad (n=0, \pm 1, \pm 2, \cdots)$$
and
$$\text{Log}(1-\sqrt{2})=\ln(\sqrt{2}-1)+(2n+1)\pi i \quad (n=0, \pm 1, \pm 2, \cdots)$$
Since $\ln(\sqrt{2}-1)=\ln\dfrac{1}{1+\sqrt{2}}=-\ln(1+\sqrt{2})$,

Thus
$$\text{Arcsin}(-i)=n\pi+i(-1)^{n+1}\ln(1+\sqrt{2}) \quad (n=0, \pm 1, \pm 2, \cdots)$$

**Example 18** To compute $\text{Arctan}(2i)$.

By expression (2.4.22), we have
$$\text{Arctan}(2i)=\frac{1}{2i}\text{Log}\frac{1+i2i}{1-i2i}=-\frac{i}{2}\text{Log}\left(-\frac{1}{3}\right)$$
$$=-\frac{i}{2}\left(\ln\frac{1}{3}+\pi i+2n\pi i\right)$$
$$=\left(\frac{1}{2}+n\right)\pi+i\,\frac{\ln 3}{2}$$

The derivatives of these three functions are readily obtained from the above expressions.
$$\frac{d}{dz}\sin^{-1}z=\frac{1}{(1-z^2)^{1/2}}, \quad \frac{d}{dz}\cos^{-1}z=\frac{-1}{(1-z^2)^{1/2}}, \quad \frac{d}{dz}\tan^{-1}z=\frac{1}{1+z^2}$$

The derivatives of the first two depend on the values chosen for the square roots. The derivative of the last one, does not, however, depend on the manner in which the function is made single-valued.

**Definition 2.4.5** We define the inverse hyperbolic function is the inverse function of the trigonometric function.

Inverse hyperbolic functions can be treated in a corresponding manner.

In order to define the inverse hyperbolic function Arch, we write
$$w=\text{Arch}z$$
when $z=\text{ch}w$. That is, when
$$z=\frac{e^w+e^{-w}}{2}$$
It turns out that

$$\text{Arch} z = \text{Log}[z \pm \sqrt{z^2-1}]$$

Similarly, we have

$$\text{Arsh} z = \text{Log}[z \pm \sqrt{z^2+1}]$$

and

$$\text{Arth} z = \frac{1}{2} \text{Log} \frac{1+z}{1-z}$$

### Exercises

1. Show that

   (a) $\log(-ei) = 1 - \frac{\pi}{2} i$ 　　(b) $\log(1-i) = \frac{1}{2}\ln 2 - \frac{\pi}{4} i$

2. Verify that

   (a) $\text{Log} e = 1 + 2n\pi i \, (n=0, \pm 1, \pm 2, \cdots)$

   (b) $\text{Log} i = \left(2n + \frac{1}{2}\right) \pi i \, (n=0, \pm 1, \pm 2, \cdots)$

   (c) $\text{Log}(-1+\sqrt{3}i) = \ln 2 + 2\left(n+\frac{1}{3}\right)\pi i \, (n=0, \pm 1, \pm 2, \cdots)$

3. Show that

   (a) $\log(1+i)^2 = 2\log(1+i)$ 　　(b) $\log(-1+i)^2 \neq 2\log(-1+i)$

4. Show that

   (a) $\text{Log}(i^{1/2}) = \left\{\left(n+\frac{1}{4}\right)\pi i: n=0, \pm 1, \pm 2, \cdots\right\} = \frac{1}{2}\text{Log} i$

   (b) $\text{Log}(i^2) \neq 2\text{Log} i$

5. Find all roots of the equation $\log z = i \frac{\pi}{2}$

6. Show in two ways that the function $\ln(x^2+y^2)$ is harmonic in every domain that does not contain the origin.

7. Show that
   $$\text{Re}[\log(z-1)] = \frac{1}{2}\ln[(x-1)^2+y^2] \, (z=x+iy \neq 1)$$

8. Show that

   (a) $(1+i)^i = e^{-\frac{\pi}{4}+2n\pi} e^{\frac{i}{2}\ln 2}, \, n=0, \pm 1, \pm 2, \cdots$

   (b) $(-1)^{1/\pi} = e^{(2n+1)i}, \, n=0, \pm 1, \pm 2, \cdots$

9. Find the principal value of

## Chapter II  Analytic Functions

(a) $i^i$  (b) $\left[\dfrac{e}{2}(-1-\sqrt{3}i)\right]^{3\pi i}$  (c) $(1-i)^{4i}$

10. Determine all values of
    (a) $2^i$  (b) $(-1)^{2i}$

11. Determine the real and imaginary parts of $z^z$.

12. To show that $(-1+\sqrt{3}i)^{3/2} = \pm 2\sqrt{2}$.

13. Let $c = a+bi$ be a fixed complex number, where $c \neq 0, \pm 1, \pm 2, \cdots$, and note that $i^c$ is multiple-valued. What restriction must be placed on the constant $c$ so that the values of $|i^c|$ are all the same?

14. Let $c$, $d$ and $z$ denote complex numbers, where $z \neq 0$. Prove that if all of the powers involved are principal values, then
    (a) $1/z^c = z^{-c}$  (b) $(z^c)^n = z^{cn}$ $(n=1, 2, \cdots)$
    (c) $z^c z^d = z^{c+d}$  (d) $z^c/z^d = z^{c-d}$

15. Assuming that $f'(z)$ exists, state the formula for the derivative of $c^{f(z)}$.

16. Find all the values of
    (a) $\tan^{-1}(2i)$  (b) $\tan^{-1}(1+i)$
    (c) $\text{ch}^{-1}(-1)$  (d) $\text{th}^{-1}0$

# Chapter III
# Complex Integration

Integrals are extremely important in the study of functions of a complex variable. The theory of integration is to be developed in this chapter.

## 1 The Concept of Contour Integrals

### 1.1 Integral of a Complex Function over a Real Interval

**Definition 3.1.1** If $f(t)$ is a complex-valued continuous function of a real variable $t$ and is written
$$f(t)=u(t)+iv(t) \tag{3.1.1}$$
where $u(t)$ and $v(t)$ are real-valued, the definite integral of $f(t)$ over an interval $[a, b]$ is defined as
$$\int_a^b f(t)\,dt = \int_a^b u(t)\,dt + i\int_a^b v(t)\,dt \tag{3.1.2}$$

**Example 1** To compute the integral
$$\int_0^1 (1+it)^2\,dt$$

By the Definition 3.1.1, we have
$$\int_0^1 (1+it)^2\,dt = \int_0^1 [(1-t^2)+i2t]\,dt = \int_0^1 (1-t^2)\,dt + i\int_0^1 2t\,dt = \frac{2}{3}+i$$

This integral has most of the properties of the real integral.

**Theorem 3.1.1** If $c=\alpha+i\beta$ is a complex constant, we have
$$\int_a^b cf(t)\,dt = c\int_a^b f(t)\,dt \tag{3.1.3}$$

**Theorem 3.1.2** If $c$ is another real number, we have
$$\int_a^b f(t)\,dt = \int_a^c f(t)\,dt + \int_c^b f(t)\,dt \tag{3.1.4}$$

**Theorem 3.1.3** When $a \leqslant b$, the inequality

$$\left|\int_a^b f(t)\,\mathrm{d}t\right| \leqslant \int_a^b |f(t)|\,\mathrm{d}t \tag{3.1.5}$$

holds for arbitrary complex function $f(t)$.

**Theorem 3.1.4** Suppose that the functions
$$f(t)=u(t)+iv(t) \quad \text{and} \quad F(t)=U(t)+iV(t)$$
are continuous on the interval $[a, b]$. If $F'(t)=f(t)$ when $t\in[a, b]$, then $U'(t)=u(t)$ and $V'(t)=v(t)$. Hence, we obtain that

$$\int_a^b f(t)\,\mathrm{d}t = U(t)\big|_a^b + iV(t)\big|_a^b = F(t)\big|_a^b = F(b) - F(a) \tag{3.1.6}$$

This theorem is the same as the fundamental theorem of calculus.

**Example 2** To compute the integral
$$\int_0^{\frac{\pi}{4}} e^{it}\,\mathrm{d}t$$

By the Theorem 3.1.4, we have
$$\int_0^{\frac{\pi}{4}} e^{it}\,\mathrm{d}t = -ie^{it}\big|_0^{\frac{\pi}{4}} = -ie^{i\frac{\pi}{4}} + i = -i\left(\frac{1}{\sqrt{2}} + \frac{i}{\sqrt{2}}\right) + i = \frac{1}{\sqrt{2}} + i\left(1 - \frac{1}{\sqrt{2}}\right)$$

Improper integrals of $f(t)$ over unbounded intervals are defined in a similar way.

## 1.2 Contour Integrals

We turn now to integrals of complex-valued functions $f(z)$ of the complex variable $z$. Such an integral is defined in terms of the values $f(z)$ along a given contour $C$ extending from a point $z=z_1$ to a point $z=z_2$ in the complex plane.

**Definition 3.1.2** Suppose that the equation
$$z=z(t)\,(a\leqslant t\leqslant b) \tag{3.1.7}$$
represents a contour $C$, extending from a point $z_1=z(a)$ to a point $z_2=z(b)$. Let the function $f(z)$ be piecewise continuous on $C$. We define the line integral, or contour integral, of $f(z)$ along $C$ to be

$$\int_C f(z)\,\mathrm{d}z = \int_a^b f[z(t)]z'(t)\,\mathrm{d}t \tag{3.1.8}$$

Note that, since $C$ is a contour, $z'(t)$ is also piecewise continuous on the interval $[a, b]$; and so the existence of integral (3.1.8) is ensured.

**Theorem 3.1.5** The integral (3.1.8) is independent of the parametrization.

**Proof** Suppose that $t=t(\tau)$ which maps an interval $\alpha\leqslant\tau\leqslant\beta$ onto $a\leqslant t\leqslant b$;

we assume that $t(\tau)$ is piecewise differentiable. By the rule for changing the variable of integration we have

$$\int_a^b f[z(t)]z'(t)dt = \int_a^\beta f\{z[t(\tau)]\}z'[t(\tau)]t'(\tau)d\tau$$

But $z'(t(\tau))t'(\tau)$ is the derivative of $z(t(\tau))$ with respect to $\tau$, and hence the integral (3. 1. 8) has the same value whether $C$ be represented by the equation $z=z(t)$ or by the equation $z=z(t(\tau))$.

**Theorem 3. 1. 6**  For any complex constant $z_0$, we have

$$\int_C z_0 f(z)dz = z_0 \int_C f(z)dz \qquad (3.1.9)$$

**Theorem 3. 1. 7**  Suppose that the complex-valued functions $f(z)$, $g(z)$ of the complex variable $z$ is defined along a given contour $C$, then

$$\int_C [f(z)+g(z)]dz = \int_C f(z)dz + \int_C g(z)dz \qquad (3.1.10)$$

provided the integrals on the right-hand sides exist.

The Theorem 3. 1. 6 and Theorem 3. 1. 7 follows immediately from Definition 3. 1. 2 and properties of integrals of complex-valued functions $f(t)$ mentioned above.

**Definition 3. 1. 3**  Suppose $C$ is the given contour with representation (3. 1. 7). Then the contour $-C$ consisting of the same set of points but with the order reversed so that the new contour extends from the point $z_2$ to the point $z_1$. The contour $-C$ has parametric representation

$$z=z(-t) \quad (-b \leqslant t \leqslant -a)$$

**Theorem 3. 1. 8**

$$\int_{-C} f(z)dz = -\int_C f(z)dz \qquad (3.1.11)$$

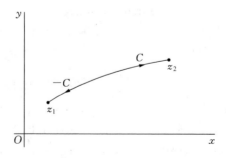

Fig. 19

Sometimes the contour $C$ is called the sum of its legs $C_1$ and $C_2$ and is denoted by $C_1+C_2$. The sum of two contours $C_1$ and $-C_2$ is well defined when $C_1$ and $C_2$ has the same final points, and it is written $C_1-C_2$.

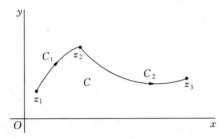

Fig. 20

**Theorem 3.1.9** Consider now a path $C$, with representation (3.1.7), that consists of a contour $C_1$ from $z_1$ to $z_2$ followed by a contour $C_2$ from $z_2$ to $z_3$, the initial point of $C_2$ being the final point of $C_1$. There is a value $c$ of $t$, where $a<c<b$, such that $z(c)=z_2$. Consequently, $C_1$ is represented by
$$z=z(t) \quad (a\leqslant t\leqslant c)$$
and $C_2$ is represented by
$$z=z(t) \quad (c\leqslant t\leqslant b)$$
Thus,
$$\int_C f(z)\,dz = \int_{C_1} f(z)\,dz + \int_{C_2} f(z)\,dz \qquad (3.1.12)$$

**Example 3** Prove that
$$\int_{|z-a|=r} \frac{dz}{(z-a)^n} = \begin{cases} 2\pi i \cdots\cdots n=1 \\ 0 \cdots\cdots\cdots n\neq 1 \end{cases}$$

**Proof** When the point $z$ is on a circle $|z-a|=r$, it follows that $z-a=re^{i\theta}$ ($0\leqslant\theta\leqslant 2\pi$), then
$$\int_{|z-a|=r} \frac{dz}{(z-a)^n} = \int_0^{2\pi} \frac{ire^{i\theta}d\theta}{(re^{i\theta})^n} = i\int_0^{2\pi} \frac{d\theta}{r^{n-1}e^{i(n-1)\theta}} = ir^{1-n}\int_0^{2\pi} e^{i(1-n)\theta}d\theta$$
$$= \begin{cases} 2\pi i \cdots\cdots n=1 \\ 0 \cdots\cdots\cdots n\neq 1 \end{cases}$$

**Example 4** Find the integral of
$$I = \int_C \bar{z}\,dz$$
from $z=-2i$ to $z=2i$ taken along a semicircle (Fig. 21).

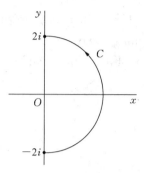

Fig. 21

$$C: z = 2e^{i\theta} \left(-\frac{\pi}{2} \leq \theta \leq \frac{\pi}{2}\right)$$

According to definition of integral

$$I = \int_{-\pi/2}^{\pi/2} \overline{2e^{i\theta}} (2e^{i\theta})' d\theta$$

Since

$$\overline{e^{i\theta}} = e^{-i\theta} \quad \text{and} \quad (e^{i\theta})' = ie^{i\theta}$$

this means that

$$I = \int_{-\pi/2}^{\pi/2} 2e^{-i\theta} 2ie^{i\theta} d\theta = 4i \int_{-\pi/2}^{\pi/2} d\theta = 4\pi i$$

**Example 5** To find the value of the integral

$$\int_C \operatorname{Re} z \, dz$$

(1) from the point 0 to the point $i$, taken along a straight line segment, and from $i$ to $1+i$;

(2) from the point 0 to the point $1+i$, taken along a straight line segment (Fig. 22).

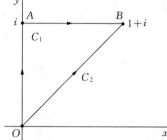

Fig. 22

The leg $OA$ may be represented parametrically as $z=0+iy(0\leqslant y\leqslant 1)$; and $f(z)=0$. Consequently,

$$\int_{OA} f(z)dz = \int_0^1 0idy = 0$$

The leg $AB$ may be represented parametrically as $z=x+i(0\leqslant x\leqslant 1)$; and $f(z)=x(0\leqslant x\leqslant 1)$, so

$$\int_{AB} f(z)dz = \int_0^1 xdx = \frac{1}{2}$$

We now see that

$$\int_{C_1} f(z)dz = \int_{OA} f(z)dz + \int_{AB} f(z)dz = \frac{1}{2}$$

The segment $OB$ with parametric representation $z=x+ix(0\leqslant x\leqslant 1)$, and $f(z)=x(0\leqslant x\leqslant 1)$. Thus

$$\int_{C_2} f(z)dz = \int_0^1 x(1+i)dx = (1+i)\int_0^1 xdx = \frac{1+i}{2}$$

Evidently, then, the integrals of $f(z)$ along the two paths $C_1$ and $C_2$ have different values even though those paths have the same initial and the same final points.

**Theroem 3.1.10** Suppose that a function $f(z)$ on a contour $C$ whose equation is $z=z(t)(a\leqslant t\leqslant b)$ and that $f(z)$ is continuous on $C$. Then for any nonnegative constant $M$ such that the values of $f(z)$ on $C$ satisfy the inequality $|f(z)|\leqslant M$, we have

$$\left|\int_C f(z)dz\right| \leqslant ML \qquad (3.1.13)$$

where $L$ denotes the length of $C$.

**Proof** We know from Definition 3.1.2, and inequality (3.1.5) that

$$\left|\int_C f(z)dz\right| = \left|\int_a^b f[z(t)]z'(t)dt\right| \leqslant \int_a^b |f[z(t)]||z'(t)|dt$$

So, for any nonnegative constant $M$ such that the values of $f(z)$ on $C$ satisfy the inequality $|f(z)|\leqslant M$, we have

$$\left|\int_C f(z)dz\right| \leqslant M\int_a^b |z'(t)|dt = ML$$

It follows that the modulus of the value of the integral of $f(z)$ along $C$ does not exceed $ML$. This completes the proof.

**Example 6** Without evaluating the integral, show that

$$\left|\int_{|z|=r}\frac{dz}{(z-a)(z+a)}\right|\leqslant\frac{2\pi r}{|r^2-|a|^2|}(r>0,|a|\neq r)$$

**Proof** If $a=0$, the integral is

$$\int_{|z|=r}\frac{dz}{z^2}=0$$

If $a\neq 0$, because $z$ is a point on the circle $|z|=2$, then
$$|(z-a)(z+a)|=|z^2-a^2|\geqslant||z|^2-a^2|=|r^2-|a|^2|$$
Thus

$$\left|\int_{|z|=r}\frac{dz}{(z-a)(z+a)}\right|\leqslant\int_{|z|=r}\left|\frac{dz}{(z-a)(z+a)}\right|=$$

$$\int_{|z|=r}\frac{|dz|}{|z^2-a^2|}\leqslant\frac{2\pi r}{|r^2-|a|^2|}(r>0,|a|\neq r)$$

## Exercises

1. Let $w(t)=u(t)+iv(t)$ denote a continuous complex-valued function defined on an interval $[-a, a]$.

   (a) Suppose that $w$ is even; that is, $w(t)=-w(t)$ for each point $t$ in the given interval. Show that
   $$\int_{-a}^{a}w(t)dt=2\int_{0}^{a}w(t)dt$$
   (b) Show that if $w$ is an odd function, that is, $w(-t)=-w(t)$ for each point $t$ in the interval, then
   $$\int_{-a}^{a}w(t)dt=0$$

2. Prove the following integration by parts formula. Let $f(t)$ and $g(t)$ be analytic in $D$ and let $\gamma$ be a rectifiable curve from $a$ to $b$ in $D$. Then
$$\int_{\gamma}f(t)g'(t)dt=f(b)g(b)-f(a)g(a)-\int_{\gamma}f'(t)g(t)dt$$

3. To evaluate

   (a) $\int_{1}^{2}\left(\frac{1}{t}-i\right)^2 dt$    (b) $\int_{0}^{\frac{\pi}{6}}e^{i2t}dt$    (c) $\int_{0}^{\infty}e^{-zt}dt\,(\text{Re}z>0)$

4. Compute
$$\int_{C}xdz$$
where $C$ is the directed line segment from 0 to $1+i$.

## Chapter III  Complex Integration

5. Compute
$$\int_{|z|=r} x\,dz$$
for the positive sense of the circle, in two ways: first, by use of a parameter, and second, by observing that $x=\frac{1}{2}(z+\bar{z})=\frac{1}{2}\left(z+\frac{r^2}{z}\right)$ on the circle.

6. Evaluate the integrals $\int_C (x-y+ix^2)\,dz$, where $C$ is the segment from 0 to $1+i$.

7. To evaluate
$$\int_C f(z)\,dz$$
$f(z)=|z|$ and $C$ is the arc from $z=-1$ to $z=1$ consisting of
   (a) the segment $-1\leqslant x\leqslant 1$ of the real axis;
   (b) the semicircle $z=e^{i\theta}$ $(0\leqslant\theta\leqslant\pi)$;
   (c) the semicircle $z=e^{i\theta}$ $(\pi\leqslant\theta\leqslant 2\pi)$.

8. To evaluate
$$\int_C f(z)\,dz$$
$f(z)=\frac{z+2}{z}$ and $C$ is
   (a) the semicircle $z=2e^{i\theta}$ $(0\leqslant\theta\leqslant\pi)$;
   (b) the semicircle $z=2e^{i\theta}$ $(\pi\leqslant\theta\leqslant 2\pi)$;
   (c) the circle $z=2e^{i\theta}$ $(0\leqslant\theta\leqslant 2\pi)$.

9. To evaluate
$$\int_C f(z)\,dz$$
$f(z)=z-1$ and $C$ is the arc from $z=0$ to $z=2$ consisting of
   (a) the semicircle $z=1+e^{i\theta}$ $(\pi\leqslant\theta\leqslant 2\pi)$;
   (b) the segment $0\leqslant x\leqslant 2$ of the real axis.

10. Suppose $f(z)=\pi e^{\pi\bar{z}}$ and $C$ is the boundary of the square with vertices at the points 0, 1, $1+i$, and $i$, the orientation of $C$ being in the counterclockwise direction.

11. Find the integral of $f(z)=e^z$ from $-3$ to 3 taken along a semicircle. Is

this integral different from the integral taken over the line segment between the two points?

12. Find the integral
$$\int_C z e^{z^2} dz$$
(a) from the point 0 to the point $2-i$, taken along a straight line segment, and

(b) from 0 to $1+i$ along the parabola $y=x^2$.

13. Find the integral
$$\int_C \sin z \, dz$$
from the origin to the point $1+i$, taken along the parabola $y=x^2$.

14. Let $C_1$ and $C_2$ be the two polygons $[1, i]$ and $[1, 1+i, i]$. Express $C_1$ and $C_2$ as paths and calculate $\int_{C_1} f(z)dz$ and $\int_{C_1} f(z)dz$ where $f(z) = |z|^2$.

15. In this question, first let $C_1$ denote the contour $OAB$, second let $C_2$ denotes the segment $OB$ of the line $y=x$, and evaluate the integral
$$\int_C f(z)dz$$
where
$$f(z)=y-x-i3x^2 \quad (z=x+iy)$$

16. Without evaluating the integral, show taht

(a) $\left| \int_C (x^2+iy^2)dz \right| \leqslant 2$, where $C$ from the point $-i$ to the point $i$, taken along a straight line segment;

(b) $\left| \int_C (x^2+iy^2)dz \right| \leqslant \pi$, where $C$ is the right-hand half of the circle, from $-i$ to $i$.

17. Let $C$ be the arc of the circle $|z|=2$ from $z=2$ to $z=2i$ that lies in the first quadrant. To show that
$$\left| \int_C \frac{z+4}{z^3-1} dz \right| \leqslant \frac{6\pi}{7}$$

18. Let $C$ be the arc of the circle $|z|=2$ from $z=2$ to $z=2i$ that lies in the first quadrant. Without evaluating the integral, show taht

## Chapter III  Complex Integration

$$\left|\int_C \frac{dz}{z^2-1}dz\right| \leq \frac{\pi}{3}$$

19. Let $C$ denote the line segment from $z=i$ to $z=1$. To show that

$$\left|\int_C \frac{dz}{z^4}\right| \leq 4\sqrt{2}$$

without evaluating the integral.

20. Show that if $C$ is the boundary of the triangle with vertices at the points $0$, $3i$, and $-4$, oriented in the counterclockwise direction, then

$$\left|\int_C (e^z - \bar{z})dz\right| \leq 60$$

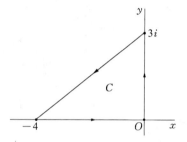

21. Let $C_R$ denote the upper half of the circle $|z|=R(R>2)$, taken in the counterclockwise direction. Show that

$$\left|\int_{C_R} \frac{2z^2-1}{z^4+5z^2+4}dz\right| \leq \frac{\pi R(2R^2+1)}{(R^2-1)(R^2-4)}$$

Then, show that the value of the integral tends to zero as $R$ tends to infinity.

22. Compute

$$\int_{|z|=2} \frac{1}{z^2-1}dz$$

for the positive sense of the circle.

23. Compute

$$\int_{|z|=1} |z-1||dz|$$

24. Suppose that $f(z)$ is analytic on a closed curve $C$ (i. e., $f(z)$ is analytic in a region that contains $C$). Show that

$$\int_C \overline{f(z)} f'(z) dz$$

is purely imaginary (The continuity of $f'(z)$ is taken for granted).

25. Assume that $f(z)$ is analytic and satisfies the inequality $|f(z)-1|<1$ in a region $D$. Show that
$$\int_C \frac{f'(z)}{f(z)}dz = 0$$
for every closed curve in $D$.

26. If $p(z)$ is a polynomial and $C$ denotes the circle $|z-a|=R$, what is the value of $\int_C p(z)d\bar{z}$ ?

27. Describe a set of circumstances under which the formula
$$\int_C \log z\, dz = 0$$
is meaningful and true.

28. Compute
$$\int_{|z|=r} \frac{|dz|}{|z-a|^2}$$
Under the condition $|a|\neq r$.

29. Let $I(r) = \int_C \frac{e^{iz}}{z}dz$ where $C: z(t)=re^{it}$, $0\leqslant t\leqslant \pi$. Show that $\lim_{r\to\infty} I(r)=0$.

## 2  Cauchy-Goursat Theorem

### 2.1  Cauchy Theorem

In Sec1, we saw that some continuous function $f(z)$, the integral of $f(z)$ along any given closed contour $C$ lying entirely in $D$ has value zero. An important class of integrals is characterized by the property that the integral along a contour depends only on its end (or final) points. In other words, if $C_1$ and $C_2$ have the same initial point and the same end point, to say that an integral deponds only on the end points is equivalent to saying that the integral over any closed contour is zero. Also we say integration is independent of path in $D$.

In this section, we present a theorem, which ensure that the value of the integral of $f(z)$ along a simple closed contour is zero. There are several forms of Cauchy's theorem.

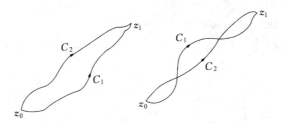

Fig. 23

**Thereom 3.2.1 (Cauchy)** If $f(z)$ is analytic on $\overline{C}$ and $f'(z)$ is continuous there, then

$$\int_C f(z)dz = 0$$

This result was obtained by Cauchy in the early part of the nineteenth century.

Goursat was the first to prove that the condition of continuity on $f'(z)$ can be omitted. Its removal is important and will allow us to show, for example, that the derivative $f'(z)$ of an analytic function $f(z)$ is analytic without having to assume the continuity $f'(z)$, which follows as a consequence. We now state the revised form of Cauchy's result, known as the Cauchy-Goursat theorem. The proof is omitted.

**Theorem 3.2.2 (Cauchy-Goursat)** If a function $f(z)$ is analytic at all points interior to and on a simple closed contour $C$, then

$$\int_C f(z)dz = 0 \qquad (3.2.1)$$

**Example 7** If $C$ is any simple closed contour, in either direction, then

$$\int_C e^{z^3} dz = 0$$

This is because the function $f(z) = e^{z^3}$ is analytic everywhere.

**Example 8** If $C$ is the unit circle $|z| = 1$, then

$$\int_C (\sin z + e^z \cos z) dz = 0$$

This is because the integrand $f(z) = \sin z + e^z \cos z$ is analytic on the closed unit disk $|z| \leqslant 1$.

**Definition 3.2.1** A simply connected domain $D$ is a domain such that every simple closed contour within it encloses only points of $D$. A domain that is not simply connected is said to the multiply connected.

In other words, a region is simply connected if a region without holes.

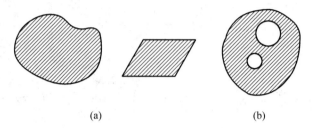

(a)                                    (b)

Fig. 24

The Cauchy-Goursat theorem can be extended in the following way, involving a simply connected domain

**Theorem 3. 2. 3** If a function $f(z)$ is analytic throughout a simply connected domain $D$, then

$$\int_C f(z)\,dz = 0$$

for every closed contour $C$ lying in $D$ (Fig. 25).

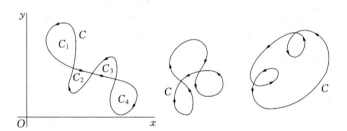

Fig. 25

**Definition 3. 2. 2** Suppose that $f(z)$ is a continuous complex function in a domain $D$. We call a function $F$ is an antiderivative of $f(z)$, such that $F'(z)=f(z)$ for all $z$ in $D$, and we can define the function

$$F(z) = \int_{z_0}^{z} f(s)\,ds \qquad (3.2.2)$$

**Corollary 3. 2. 4** A function $f(z)$ that is analytic throughout a simply connected domain $D$ must have an antiderivative everywhere in $D$.

**Corollary 3.2.5** The integral $\int_C f(z)dz$, with continuous function $f(z)$, depends only on the end points of $C$ if and only if $f(z)$ is the derivative of an analytic function in $D$.

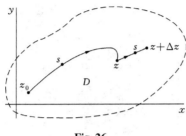

Fig. 26

**Example 9** To find the integral

$$\int_{2+3i}^{1-i} z^3 dz$$

The continuous function $f(z) = z^3$ has an antiderivative $F(z) = \dfrac{z^4}{4}$ throughout the plane. Hence

$$\int_{2+3i}^{1-i} z^3 dz = \dfrac{z^4}{4}\bigg|_{2+3i}^{1-i} = \dfrac{(1-i)^4}{4} - \dfrac{(2+3i)^4}{4}$$

for every contour from $z=0$ to $z=1+i$.

The Cauchy-Goursat theorem can also be extended in a way that involves integrals along the boundary of a multiply connected domain. The following theorem is such an extension, called extended Cauchy-Goursat Theorem.

**Theorem 3.2.6** Suppose that

(1) $C$ is a simple closed contour, described in the counterclockwise direction;

(2) $C_k(k=1, 2, \cdots, n)$ are simple closed contours interior to $C$, all described in the clockwise direction, that are disjoint and whose interiors have no points in common (Fig. 27).

If a function $f(z)$ is analytic on all of these contours and throughout the multiply connected domain consisting of all points inside $C$ and exterior to each $C_k$, then

$$\int_C f(z)dz + \sum_{k=1}^{n} \int_{C_k^-} f(z)dz = 0 \qquad (3.2.3)$$

or

$$\int_C f(z)dz = \sum_{k=1}^{n} \int_{C_k} f(z)dz \qquad (3.2.4)$$

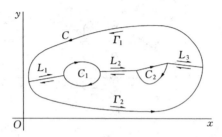

Fig. 27

## 2.2  Cauchy Integral Formula

**Theorem 3.2.7(Cauchy Integral Formula)**   Let $f(z)$ be analytic everywhere inside and on a simple closed contour $C$, taken in the positive sense. If $z_0$ is any point interior to $C$, then

$$f(z_0) = \frac{1}{2\pi i} \int_C \frac{f(z)dz}{z - z_0} \qquad (3.2.5)$$

**Proof**   Let $C_\rho$ be the circle of radius $\rho$ centered at $z_0$, denote a positively oriented circle $|z - z_0| = \rho$, where $\rho$ is small enough that $C_\rho$ is interior to $C$ (see Fig. 28). Since the function $\dfrac{f(z)}{z - z_0}$ is analytic between and on the contours $C$ and $C_\rho$, it follows from the Theorem 3.2.6 that

$$\int_C \frac{f(z)dz}{z - z_0} = \int_{C_\rho} \frac{f(z)dz}{z - z_0} \qquad (3.2.6)$$

This enables us to write

$$\int_C \frac{f(z)dz}{z - z_0} - 2\pi i f(z_0) = \int_{C_\rho} \frac{f(z)}{z - z_0}dz - \int_{C_\rho} \frac{f(z_0)}{z - z_0}dz$$

$$= \int_{C_\rho} \frac{f(z) - f(z_0)}{z - z_0}dz \qquad (3.2.7)$$

Chapter III   Complex Integration

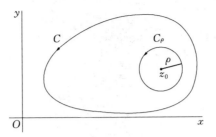

Fig. 28

Since $f(z)$ is analytic, and therefore continuous, at $z_0$, then for each positive number $\varepsilon$, there is a positive number $\delta$ such that
$$|f(z)-f(z_0)|<\varepsilon \text{ whenever } |z-z_0|<\delta \qquad (3.2.8)$$

Let the radius $\rho$ of the circle $C_\rho$ be smaller than the number $\delta$ in the second of these inequalities. Since $|z-z_0|=\rho$ when $z$ is on $C_\rho$, it follows that inequalities (3.2.8) holds when $z$ is such a point; tells us that
$$\left|\int_{C_\rho}\frac{f(z)-f(z_0)}{z-z_0}dz\right|<\frac{\varepsilon}{\rho}2\pi\rho=2\pi\varepsilon$$

In view of equation (3.2.6), then,
$$\left|\int_C\frac{f(z)dz}{z-z_0}-2\pi i f(z_0)\right|<2\pi\varepsilon$$

Since the left-hand side of this inequality is a nonnegative constant that is less than an arbitrarily small positive number, it must equal to zero. Hence equation (3.2.5) is valid, and the theorem is proved.

Formula (3.2.5) is called the Cauchy integral formula. It tells us that if a function $f(z)$ is analytic within and on a simple closed contour $C$, then the values of $f(z)$ interior to $C$ are completely determined by the values of $f(z)$ on $C$.

When the Cauchy integral formula is written
$$\int_C\frac{f(z)dz}{z-z_0}=2\pi i f(z_0) \qquad (3.2.9)$$

it can be used to evaluate certain integrals along simple closed contours.

**Example 10**   To find the integral
$$\int_C\frac{e^z}{z}dz$$

taken over a path $C: |z|=1$.

Since the function

$$f(z)=e^z$$
is analytic within and on $C$ and since the point $z_0=0$ is interior to $C$, formula (3.2.9) tells us that
$$\int_C \frac{e^z}{z}dz = 2\pi i f(0) = 2\pi i$$

**Example 11** To evaluate the integral
$$\int_C \frac{zdz}{(9-z^2)(z+i)}$$
where $C$ be the positively oriented circle $|z|=2$.

Since the function
$$f(z)=\frac{z}{9-z^2}$$
is analytic within and on $C$ and since the point $z_0=-i$ is interior to $C$, formula (3.2.9) tells us that
$$\int_C \frac{zdz}{(9-z^2)(z+i)} = \int_C \frac{z/(9-z^2)}{z-(-i)}dz = 2\pi i\left(\frac{-i}{10}\right)=\frac{\pi}{5}$$

## 2.3 Derivatives of Analytic Functions

It follows from the Cauchy integral formula that if a function is analytic at a point, then its derivatives of all orders exist at that point and are themselves analytic there.

**Theorem 3.2.8** An analytic function has derivatives of all orders which are analytic and can be represented by the formula
$$f^{(n)}(z_0) = \frac{n!}{2\pi i}\int_C \frac{f(z)dz}{(z-z_0)^{n+1}} \quad (n=1, 2, \cdots) \qquad (3.2.10)$$

When written in the form
$$\int_C \frac{f(z)dz}{(z-z_0)^{n+1}} = \frac{2\pi i}{n!}f^{(n)}(z_0)(n=1, 2, \cdots) \qquad (3.2.11)$$

Expression (3.2.11) can be useful in evaluating certain integrals when $f(z)$ is analytic inside and on a simple closed contour $C$, taken in the positive sense, and $z_0$ is any point interior to $C$.

**Corollary 3.2.9** If a function $f(z)=u(x, y)+iv(x, y)$ is defined and analytic at a point $z=(x, y)$, then the component functions $u$ and $v$ has continuous partial derivatives of all orders at that point.

**Example 12** To find the integral

$$\int_C \frac{\cos z \, dz}{(z-i)^3}$$

where $C$ is the positively oriented unit circle $|z-i|=1$.

Since the function

$$f(z) = \frac{\cos z}{(z-i)^3}$$

is analytic within and on $C$ and since the point $z_0=i$ is interior to $C$, formula (3.2.11) tells us that

$$\int_C \frac{\cos z \, dz}{(z-i)^3} = \int_C \frac{f(z) \, dz}{(z-i)^{3+1}} = \frac{2\pi i}{2!} f''(0) = -\pi i \cos i = -\pi \frac{e^{-1}+e}{2} i$$

**Theorem 3.2.10 (Morea)**  If $f(z)$ is defined and continuous in a domain $D$ and if

$$\int_C f(z) \, dz = 0 \qquad (3.2.12)$$

for all closed contour $C$ in $D$, then $f(z)$ is analytic throughout $D$.

## 2.4  Liouville's Theorem and the Fundamental Theorem of Algebra

**Theorem 3.2.11 (Liouville)**  A function which is analytic and bounded in the whole plane must reduce to a constant.

**Proof**  For each point $z_0$ in the complex plane. Let $C_R$ be the circle of radius $R$ centered at $z_0$, and assume that $|f(z)| \leq M$ for all $z$. By equation (3.2.10), we obtain at once

$$|f'(z_0)| \leq \frac{1!}{2\pi} \int_{C_R} \left| \frac{f(z) \, dz}{(z-z_0)^2} \right| \leq \frac{M}{R} \qquad (3.2.13)$$

Now the number $M$ in inequality (3.2.13) is independent of the value of $R$ that is taken. Hence that inequality can hold for arbitrarily large values of $R$ only if $f'(z_0)=0$. Since the choice of $z_0$ was arbitrary, this means that $f'(z)=0$ everywhere in the complex plane. Consequently, $f(z)$ is a constant function.

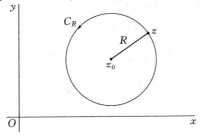

Fig. 29

The following theorem, known as the fundamental theorem of algebra, follows readily from Liouville's theorem.

**Theorem 3. 2. 12** (**Fundamental Theorem of Algebra**) Any polynomial
$$P(z)=a_0+a_1z+a_2z^2+\cdots+a_nz^n\ (a_n\neq 0)$$
of degree $n(n\geq 1)$ has at least one zero. That is, there exists at least one point $z_0$ such that $P(z_0)=0$.

**Proof** Supposed that
$$f(z)=\frac{1}{P(z)}$$
Then the function $f(z)$ would be analytic in the whole complex plane. Because $\lim_{z\to\infty}P(z)=\infty$, therefore $\lim_{z\to\infty}f(z)=0$. This implies boundedness, and by Liouville's theorem $f(z)$ would be constant. Since this is not so, the equation $P(z)=0$ must have a root.

## Exercises

1. Apply the Cauchy-Goursat theorem to show that
$$\int_C f(z)dz=0$$
When the contour is the circle $|z|=1$, in either direction, and when

   (a) $\dfrac{1}{\cos z}$      (b) $\dfrac{1}{z^2+2z+2}$      (c) $\dfrac{1}{z^2+5z+6}$

   (d) $z\cos z^2$      (e) $f(z)=\dfrac{z^2}{z-3}$      (f) $f(z)=ze^{-z}$

   (g) $f(z)=\tan z$

2. Let $C_1$ denote the positively oriented circle $|z|=4$ and $C_2$ the positively oriented boundary of the square whose sides lie along the lines $x=\pm 1$, $y\pm 1$. With the aid of Cauchy's Theorem, point out why
$$\int_{C_1}f(z)dz=\int_{C_2}f(z)dz$$
when

   (a) $f(z)=\dfrac{1}{3z^2+1}$      (b) $f(z)=\dfrac{z+2}{\sin\dfrac{z}{2}}$

   (c) $f(z)=\dfrac{z}{1-e^z}$

## Chapter III  Complex Integration

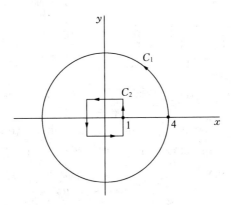

3. By finding an antiderivative, evaluate each of these integrals, where the path is any contour between the indicated limits of integration:

(a) $\int_{i}^{\frac{i}{2}} e^{\pi z} dz$    (b) $\int_{0}^{\pi+2i} \cos\left(\frac{z}{2}\right) dz$    (c) $\int_{1}^{3} (z-2)^{3} dz$

4. To evaluated the integral

$$\int_{C} \frac{\sin \frac{\pi}{4} z}{z^{2}-1} dz$$

(a) $C_1$: $|z+1|=\dfrac{1}{2}$    (b) $C_2$: $|z-1|=\dfrac{1}{2}$    (c) $C_3$: $|z|=2$

5. Let $C$ be the circle $x^2+y^2=3$, $f(z) = \displaystyle\int_{C} \frac{3\xi^{2}+7\xi+1}{\xi-z} d\xi$. To evaluate $f'(1+i)$.

6. Find the integrals over the unit circle $C$:

(a) $\displaystyle\int_{C} \frac{\cos z}{z} dz$    (b) $\displaystyle\int_{C} \frac{\sin z}{z} dz$    (c) $\displaystyle\int_{C} \frac{\cos z^{2}}{z} dz$

7. Let $C$ denote the positively oriented boundary of the square whose sides lie along the lines $x=\pm 2$ and $y=\pm 2$. Evaluate each of these integrals:

(a) $\displaystyle\int_{C} \frac{e^{-z} dz}{z-(\pi i/2)}$    (b) $\displaystyle\int_{C} \frac{\cos z}{z(z^{2}+8)} dz$    (c) $\displaystyle\int_{C} \frac{z dz}{2z+1}$

(d) $\displaystyle\int_{C} \frac{\operatorname{ch} z}{z^{4}} dz$    (e) $\displaystyle\int_{C} \frac{\tan \frac{z}{2}}{(z-x_{0})^{2}} dx (-2 < x_0 < 2)$

8. Find the value of the integral of $g(z)$ around the circle $|z-i|=2$ in the positive sense

when

(a) $g(z) = \dfrac{1}{z^2+4}$    (b) $g(z) = \dfrac{1}{(z^2+4)^2}$

9. Let $C$ be the circle $|z|=3$, described in the positive sense. Show that if
$$g(w) = \int_C \frac{2z^2-z-2}{z-w}dz \quad (|w|\neq 3)$$
then $g(2)=8\pi i$. What is the value of $g(w)$ when $|w|>3$?

10. Let $C$ be any simple closed contour, described in the positive sense in the $z$ plane, and write
$$g(w) = \int_C \frac{z^3+2z}{(z-w)^3}dz$$
Show that $g(w)=6\pi i w$ when $w$ is inside $C$ and that $g(w)=0$ when $w$ is outside $C$.

11. Let $C$ be the unit circle $z=e^{i\theta}$ $(-\pi \leqslant \theta \leqslant \pi)$. First show that, for any real constant $a$,
$$\int_C \frac{e^{az}}{z}dz = 2\pi i$$
Then write this integral in terms of $\theta$ to derive the integration formula
$$\int_0^\pi e^{a\cos\theta}\cos(a\sin\theta)d\theta = \pi$$

12. Prove that a function which is analytic in the whole plane and satisfies an inequality $|f(z)|<|z|^n$ for some $n$ and all sufficiently large $|z|$ reduces to a polynomial.

13. If $f(z)$ is analytic and $|f(z)|\leqslant M$ for $|z|\leqslant R$, find an upper bound for $|f^{(n)}(z)|$ in $|z|\leqslant r<R$.

14. If $f(z)$ is analytic for $|z|\leqslant 1$ and $|f(z)|\leqslant \dfrac{1}{1-|z|}$, find the best estimate of $|f^{(n)}(0)|$ that Cauchy's inequality will yield.

15. Show that the successive derivatives of an analytic function at a point can never satisfy $|f^{(n)}(z)|>n!\, n^n$. Formulate a sharper theorem of the same kind.

16. Let $p(z)$ be a polynomial of degree $n$ and let $R>0$ be sufficiently large so that $p(z)$ never vanishes in $\{z: |z|>R\}$. If $\gamma(t)=Re^{it}$, $0\leqslant t\leqslant 2\pi$, show that

Chapter III  Complex Integration

$$\int_\gamma \frac{p'(z)}{p(z)} dz = 2n\pi i$$

## 3  Harmonic Functions

**Definition 3.3.1**  A real-valued function $u(x, y)$ of two real variables $x$ and $y$, defined and single-valued in a given domain of the $xy$ plane, is said to be harmonic if, throughout that domain, it has continuous partial derivatives of the first and second order and satisfies the partial differential equation

$$u_{xx}(x, y) + u_{yy}(x, y) = 0 \tag{3.3.1}$$

known as Laplace's equation.

We usually define the Laplace's operator

$$\Delta = \left(\frac{\partial}{\partial x}\right)^2 + \left(\frac{\partial}{\partial y}\right)^2 \tag{3.3.2}$$

And so $u(x, y)$ is harmonic if and only if $\Delta u = 0$.

**Theorem 3.3.1**  The sum of two harmonic functions and a constant multiple of a harmonic function are also harmonic function.

This is due to the linear character of Laplace's equation.

**Theorem 3.3.2**  If a function $f(z) = u(x, y) + iv(x, y)$ is analytic in a domain $D$, then its component functions $u$ and $v$ are harmonic in $D$.

**Proof**  Assuming that $f(z)$ is analytic in $D$, by the Cauchy-Riemann equations

$$u_x = v_y, \quad u_y = -v_x \tag{3.3.3}$$

Differentiating both sides of these equations with respect to $x$, we have

$$u_{xx} = v_{yx}, \quad u_{yx} = -v_{xx} \tag{3.3.4}$$

Likewise, differentiation with respect to $y$ yields

$$u_{xy} = v_{yy}, \quad u_{yy} = -v_{xy} \tag{3.3.5}$$

Now, by a theorem in calculus, the continuity of the partial derivatives of $u$ and $v$ ensures that $u_{yx} = u_{xy}$ and $v_{yx} = v_{xy}$. It then follows from equations (3.3.4) and (3.3.5) that

$$u_{xx} + u_{yy} = 0 \text{ and } v_{xx} + v_{yy} = 0$$

That is, $u$ and $v$ are harmonic in $D$.

**Example 13**  Since the function $f(z) = \frac{i}{z^2}$ is analytic whenever $z \neq 0$ and since

$$\frac{i}{z^2} = \frac{i}{z^2} \cdot \frac{\bar{z}^2}{\bar{z}^2} = \frac{i\bar{z}^2}{(z\bar{z})^2} = \frac{i\bar{z}^2}{|z|^4} = \frac{2xy + i(x^2 - y^2)}{(x^2 + y^2)^2}$$

the two functions

$$u(x, y) = \frac{2xy}{(x^2 + y^2)^2} \text{ and } v(x, y) = \frac{x^2 - y^2}{(x^2 + y^2)^2}$$

are harmonic throughout any domain in the $xy$ plane that does not contain the origin.

**Definition 3.3.2** If two given functions $u$ and $v$ are harmonic in a domain $D$ and their first-order partial derivatives satisfy the Cauchy-Riemann equation throughout $D$, $v$ is said to be a harmonic conjugate of $u$.

**Theorem 3.3.3** A function $f(z) = u(x, y) + iv(x, y)$ is analytic in a domain $D$ if and only if $v$ is a harmonic conjugate of $u$.

**Example 14** Suppose that the function

$$f(z) = z^2$$

Since $u(x, y) = x^2 - y^2$ and $v(x, y) = 2xy$ are the real and imaginary components respectively. We know that $v$ is a harmonic conjugate of $u$ throughout the plane. But $u$ cannot be a harmonic conjugate of $v$.

**Example 15** Show that $u(x, y)$ is harmonic in the complex plane and find a harmonic conjugate $v(x, y)$ when $u(x, y) = x^3 - 3xy^2$

Since

$$u_x = 3x^2 - 3y^2, \quad u_y = -6xy$$
$$u_{xx} = 6x, \quad u_{xx} = -6x$$

So $u(x, y)$ is harmonic in the complex plane.

Since a harmonic conjugate $v(x, y)$ is related to $u(x, y)$ by means of the Cauchy-Riemann equations

$$u_x = v_y, \quad u_y = -v_x$$

the second of these equations tells us that

$$v_x(x, y) = 6xy$$

Holding $y$ fixed and integrating each side here with respect to $x$, we find that

$$v(x, y) = 3x^2 y + \phi(y)$$

where $\phi$ is an arbitrary function of $y$. We have

$$3x^2 - 3y^2 = 3x^2 + \phi'(y)$$

Thus $\phi(x) = y^3 + C$, where $C$ is an arbitrary real number. Then, the function

Chapter III  Complex Integration

$$v(x, y) = 3x^2 y - y^3 + C$$

is a harmonic conjugate of $u(x, y)$.

The corresponding analytic function is

$$f(z) = (x^3 - 3xy^2) + i(3x^2 y - y^3 + C) = (x+iy)^3 + iC = z^3 + iC$$

### Exercises

1. Show that $u(x, y)$ is harmonic in some domain and find a harmonic conjugate $v(x, y)$ when
   (a) $u(x, y) = x^2 + xy - y^2$      $f(i) = -1 + i$
   (b) $u(x, y) = e^x(x\cos y - y\sin y)$      $f(0) = 0$
   (c) $v(x, y) = \dfrac{y}{x^2 + y^2}$      $f(2) = 0$

2. Show that if $v$ and $V$ are harmonic conjugates of $u$ in a domain $D$. Then $v(x, y)$ and $V(x, y)$ can differ at most by an additive constant.

3. Suppose that, in a domain $D$, a function $v$ is a harmonic conjugate of $u$ and also that $u$ is a harmonic conjugate of $v$. Show how it follows that both $u(x, y)$ and $v(x, y)$ must be constant throughout $D$.

4. To show that, in a domain $D$, $v$ is a harmonic conjugate of $u$ if and only if $-u$ is a harmonic conjugate of $v$.

5. Prove that the functions $u(z)$ and $\overline{u(z)}$ are simultaneously harmonic.

6. Let the function $f(z) = u(x, y) + iv(x, y)$ be analytic in some domain $D$. State why the functions

$$U(x,y) = e^{u(x,y)} \cos v(x,y) ; V(x,y) = e^{u(x,y)} \sin v(x,y)$$

are harmonic in $D$ and why $V(x, y)$ is, in fact, a harmonic conjugate of $U(x, y)$.

7. Let $u(z)$ be a harmonic function such that $u(z) \geq 0$ for all $z$; prove that $u(z)$ is constant.

8. If $u(z)$ is harmonic function, show that $f(z) = u_x - iu_y$ is analytic.

9. Show that if $u(z)$ is harmonic function then so are $u_x$ and $u_y$.

10. For $|z| < 1$ let

$$u(z) = \mathrm{Im}\left[\left(\frac{1+z}{1-z}\right)^2\right]$$

Show that $u(z)$ is harmonic and $\lim_{r \to 1^-} u(re^{i\theta}) = 0$.

# Chapter IV

## Series

This chapter is devoted mainly to power series representations of analytic functions. Similarly, we shall extend usual series to complex valued functions, and we shall see that they have similar properties to the functions of a real variable which you have already known.

## 1  Basic Properties of Series

### 1.1  Convergence of Sequences

**Definition 4.1.1**  If $z_n$ is complex number for every positive integer $n$ then the infinite sequence

$$\{z_n\}, \text{ i.e., } z_1, z_2, \cdots, z_n, \cdots \tag{4.1.1}$$

of complex numbers has a limit $z$ if for every positive number $\varepsilon$, there exists a positive integer $N$ such that

$$|z_n - z| < \varepsilon \text{ whenever } n > N \tag{4.1.2}$$

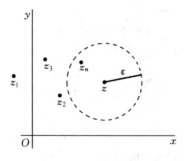

Fig. 30

When that limit exists, the sequence is said to converge to $z$; and we write

$$\lim_{n \to \infty} z_n = z, \text{ or simply}, z_n \to z(n \to \infty) \tag{4.1.3}$$

If the limit of a sequence $\{z_n\}$ does not exist, then we call the sequence $\{z_n\}$ divergent. If $\lim\limits_{n\to\infty} z_n = \infty$, the sequence may be said to diverge to infinity.

We note that all the usual theorems about limits holds for complex numbers: limits of sums, limits of products, limits of quotients, limits of composite functions. The proofs which you had in calculus hold without change in the present context. It is then usually easy to compute limits.

**Example 1** Find the limit
$$\lim_{n\to\infty} \frac{nz}{1+nz}$$
for any complex number $z$.

If $z=0$, it is clear that the limit is zero. Suppose $z \neq 0$, then the limit is
$$\lim_{n\to\infty} \frac{nz}{1+nz} = \lim_{n\to\infty} \frac{z}{\frac{1}{n}+z} = 1$$

**Theorem 4.1.1** Suppose that $z_n = x_n + iy_n$ ($n = 1, 2, \cdots$) and $z = x+iy$. Then
$$\lim_{n\to\infty} z_n = z \tag{4.1.4}$$
if and only if
$$\lim_{n\to\infty} x_n = x \quad \text{and} \quad \lim_{n\to\infty} y_n = y \tag{4.1.5}$$

**Proof** We first assume the conditions (4.1.5) hold.

According to conditions (4.1.5), for each positive number $\varepsilon$, there exist positive integers $n_1$ and $n_2$ such that
$$|x_n - x| < \frac{\varepsilon}{2} \quad \text{whenever} \quad n > n_1$$
and
$$|y_n - y| < \frac{\varepsilon}{2} \quad \text{whenever} \quad n > n_2$$

Hence, if $n_0$ is the larger of the two integers $n_1$ and $n_2$,
$$|x_n - x| < \frac{\varepsilon}{2} \quad \text{and} \quad |y_n - y| < \frac{\varepsilon}{2} \quad \text{whenever} \quad n > n_0$$

Since
$$|(x_n+iy_n) - (x+iy)| = |(x_n-x) + i(y_n-y)| \leqslant |x_n-x| + |y_n-y|$$
then,

$$|z_n - z| < \frac{\varepsilon}{2} + \frac{\varepsilon}{2} = \varepsilon \quad \text{whenever} \quad n > n_0$$

Conversely, if we start with condition (4.1.4), we know that, for each positive number $\varepsilon$, there exists a positive integer $n_0$ such that
$$|(x_n + iy_n) - (x + iy)| < \varepsilon \quad \text{whenever} \quad n > n_0$$
But
$$|x_n - x| \leq |(x_n - x) + i(y_n - y)| = |(x_n + iy_n) - (x + iy)|$$
and
$$|y_n - y| \leq |(x_n - x) + i(y_n - y)| = |(x_n + iy_n) - (x + iy)|$$
$$|x_n - x| < \varepsilon \quad \text{and} \quad |y_n - y| < \varepsilon \quad \text{whenever} \quad n > n_0$$
Thus, conditions (4.1.5) are satisfied. This completes the proof.

Theorem 4.1.1 enables us to write
$$\lim_{n \to \infty}(x_n + iy_n) = \lim_{n \to \infty} x_n + i \lim_{n \to \infty} y_n$$

**Example 2** Find the limit
$$\lim_{n \to \infty}\left(\frac{1}{n^3} + i\right)$$

Since
$$\lim_{n \to \infty}\left(\frac{1}{n^3} + i\right) = \lim_{n \to \infty} \frac{1}{n^3} + i \lim_{n \to \infty} 1 = 0 + i \cdot 1 = i$$

We can also use Definition 4.1.1 to obtain this result. For each positive number $\varepsilon$,
$$|z_n - i| < \varepsilon \quad \text{whenever} \quad n > \frac{1}{\sqrt[3]{\varepsilon}}$$

## 1.2 Convergence of Series

**Definition 4.1.2** If $z_n$ is complex number for every positive integer $n$, then the series
$$\sum_{n=1}^{\infty} z_n = z_1 + z_2 + \cdots + z_n + \cdots \quad (4.1.6)$$
converges to $S$ if for every positive number $\varepsilon$, there exists a positive integer $m$ such that
$$\left|\sum_{n=1}^{N} z_n - S\right| < \varepsilon \quad \text{whenever} \quad N > m$$
We define the partial sum

$$S_N = \sum_{n=1}^{N} z_n = z_1 + z_2 + \cdots + z_N \quad (N = 1, 2, \cdots) \qquad (4.1.7)$$

We also say that the series converges if
$$\lim_{N \to \infty} S_N = S$$
exists, in this case we say that $S$ is equal to the sum of series, that is
$$\sum_{n=1}^{\infty} z_n = S$$

When a series does not converge, we say that it diverge.

**Theorem 4.1.2** Suppose that $z_n = x_n + iy_n$ ($n = 1, 2, \cdots$) and $S = X + iY$. Then
$$\sum_{n=1}^{\infty} z_n = S \qquad (4.1.8)$$
if and only if
$$\sum_{n=1}^{\infty} x_n = X \quad \text{and} \quad \sum_{n=1}^{\infty} y_n = Y \qquad (4.1.9)$$

**Proof** We first write (4.1.7) as
$$S_N = X_N + iY_N \qquad (4.1.10)$$
where
$$X_N = \sum_{n=1}^{N} x_n \quad \text{and} \quad Y_N = \sum_{n=1}^{N} y_n$$

Now (4.1.8) is true if and only if
$$\lim_{N \to \infty} S_N = S \qquad (4.1.11)$$
and, in view of (4.1.11) and Theorem 4.1.1, limit (4.1.11) holds if and only if
$$\lim_{N \to \infty} X_N = X \quad \text{and} \quad \lim_{N \to \infty} Y_N = Y \qquad (4.1.12)$$

Limits (4.1.9) therefore imply statement (4.1.8), and conversely.

This theorem tells us that
$$\sum_{n=1}^{\infty} (x_n + iy_n) = \sum_{n=1}^{\infty} x_n + i \sum_{n=1}^{\infty} y_n$$

**Corollary 4.1.3** If series (4.1.6) is convergence, then
$$\lim_{n \to \infty} z_n = 0 \qquad (4.1.13)$$

**Definition 4.1.3** Let $\sum_{n=1}^{\infty} z_n$ be a series of complex numbers. We shall say that this series converges absolutely if the real positive series $\sum_{n=1}^{\infty}|z_n|$ convergence.

**Theorem 4.1.4** If a series converges absolutely, then it converges.

**Theorem 4.1.5** Let $\sum_{n=1}^{\infty}a_n$ be a series of real numbers $\geq 0$ which converges. If $|z_n|\leq a_n$ for all $n$, then the series $\sum_{n=1}^{\infty}z_n$ converges absolutely.

It is often convenient to define the remainder $\rho_N$ after $N$ terms:
$$\rho_N = S - S_N \qquad (4.1.14)$$
Thus $S = S_N + \rho_N$, and, since $|S_N - S| = |\rho_N - 0|$, we see that the following conclusion holds.

**Theorem 4.1.6** A series converges to a number $S$ if and only if the sequence of remainders tends to zero.

## 1.3 Uniform convergence

**Definition 4.1.4** Suppose a sequence of functions $f_n(z)$ all defined on the same domain $D$. The sequence $\{f_n(z)\}$ converges uniformly to $f(z)$ on the domain $D$ if for every positive number $\varepsilon$, there exists a positive integer $N$ such that $|f_n(z) - f(z)| < \varepsilon$ for all $n > N$ and all $z$ in $D$.

**Theorem 4.1.7** The limit function of a uniformly convergent sequence of continuous functions is itself continuous.

**Theorem 4.1.8 (Cauchy's Necessary and Sufficient Condition)** The sequence $\{f_n(z)\}$ converges uniformly on the domain $D$ if and only if for every positive number $\varepsilon$, there exists a positive integer $N$ such that $|f_n(z) - f_m(z)| < \varepsilon$ for all $n > N$ and all $z$ in $D$.

**Theorem 4.1.9** Suppose a sequence of functions $f_n(z)$ all defined on the same domain $D$, $|f_n(z)| \leq M_n$ for all $n$ and every $z$ in $D$, and suppose the constants satisfy $\sum_{n=1}^{\infty}M_n < \infty$. Then $\sum_{n=1}^{\infty}f_n(z)$ is uniformly converges.

**Theorem 4.1.10 (Weierstrass)** Suppose that $f_n(z)$ is analytic in the region $D_n$, and that the sequence $\{f_n(z)\}$ converges to a limit function $f(z)$ in a region $D$, uniformly on every compact subset $K$ of $D$. Then $f(z)$ is analytic in $D$. Moreover, the sequence of derivatives $\{f_n'(z)\}$ converges uniformly on every compact subset $K$, and $\lim\limits_{n\to\infty} f_n'(z) = f'(z)$.

Hint: Cover the compact set with a finite number of closed discs contained in $U$, and of sufficiently small radius. Cauchy's formula expresses the derivative $f_n'(z)$ as an integral.

**Corollary 4.1.11** If a series with analytic terms
$$f(z) = f_1(z) + f_2(z) + \cdots + f_n(z) + \cdots,$$
converges uniformly on every compact subset of a region $D$, then the sum $f(z)$ is analytic in $D$, and the series can be differentiated term by term.

## Exercises

1. Let $\alpha$ be a complex number of absolute value $<1$. What is $\lim\limits_{n\to\infty} \alpha^n$? Proof?

2. If $|\alpha|>1$, does $\lim\limits_{n\to\infty} \alpha^n$ exist? Why?

3. Show in two ways that the sequence
$$z_n = -2 + i\frac{(-1)^n}{n^2} \quad (n=1, 2, \cdots)$$
converges to $-2$.

4. Prove that a convergent sequence is bounded.

5. Discuss the convergence of the sequence

   (a) $z_n = \left(1 + \dfrac{1}{n}\right) e^{i\frac{\pi}{n}}$
   
   (b) $z_n = \left(1 + \dfrac{i}{2}\right)^{-n}$
   
   (c) $z_n = (-1)^n + \dfrac{i}{n+1}$
   
   (d) $z_n = \dfrac{1}{n} e^{-i\frac{n\pi}{2}}$

6. If $\lim\limits_{n\to\infty} z_n = A$, prove that
$$\lim_{n\to\infty} \frac{z_1 + z_2 + \cdots + z_n}{n} = A$$

7. Show that for any complex number $z \neq 1$, we have
$$1 + z + z^2 + \cdots + z^n = \frac{z^{n+1} - 1}{z - 1}$$
If $|z|<1$, show that
$$\lim_{n\to\infty}(1 + z + z^2 + \cdots + z^n) = \frac{1}{1-z}$$

8. Show that

if $\lim_{n\to\infty} z_n = A$, then $\lim_{n\to\infty} |z_n| = |A|$

9. Let $f(z)$ be the function defined by
$$f(z) = \lim_{n\to\infty} \frac{1}{1+n^2 z}$$
Show that $f(0)=1$, and $f(z)=0$ if $z\neq 0$.

10. For $|z|\neq 1$ show that the following limit exists
$$f(z) = \lim_{n\to\infty}\left(\frac{z^n-1}{z^n+1}\right)$$
Is it possible to define $f(z)$ when $|z|=1$ in such way to make $f(z)$ continuous?

11. Let
$$f(z) = \lim_{n\to\infty} \frac{z^n}{1+z^n}$$
(a) What is the domain of definition of $f(z)$, that is, for which complex numbers $z$ does the limit exists?
(b) Give explicitly the value of $f(z)$ for the various $z$.

12. Discuss the convergence of the series

(a) $\sum_{n=1}^{\infty} \frac{i^n}{n}$  
(b) $\sum_{n=1}^{\infty} \frac{(3+5i)^n}{n!}$  
(c) $\sum_{n=1}^{\infty} \left(\frac{1+5i}{2}\right)^n$  
(c) $\sum_{n=2}^{\infty} \frac{i^n}{\ln n}$  
(d) $\sum_{n=1}^{\infty} \frac{1}{n}\left(1+\frac{i}{n}\right)$  
(e) $\sum_{n=1}^{\infty} \left[\frac{(-1)^n}{n} + \frac{1}{2^n}i\right]$

13. Show that
$$\text{if } \sum_{n=1}^{\infty} z_n = S, \quad \text{then} \quad \sum_{n=1}^{\infty} \overline{z_n} = \overline{S}$$

14. Show that the sum of an absolutely convergent series does not change if the terms are rearranged.

15. Discuss completely the convergence and uniform convergence of the sequence $\{nz^n\}$.

## 2 Power Series

**Definition 4. 2. 1** A power series is of the form
$$\sum_{n=0}^{\infty} c_n (z-z_0)^n \qquad (4.2.1)$$
where $z_0$, the coefficients $c_n$ and the variable $z$ are complex.

**Theorem 4.2.1(Hadamard's Formula)** Suppose the power series (4.2.1), and let $R$ be its radius of convergence. Then

$$\frac{1}{R} = \lim_{n\to\infty}\left|\frac{c_{n+1}}{c_n}\right| \quad \text{or} \quad \frac{1}{R} = \lim_{n\to\infty}\sqrt[n]{|c_n|} \quad (4.2.2)$$

**Definition 4.2.2** The number $R$ is called the radius of convergence of the power series. The circle $|z-z_0|=R$ is called the circle of convergence.

**Example 3** Determine the radius of convergence for the following power series

(a) $\sum_{n=1}^{\infty}\frac{z^n}{n^2}$ (b) $\sum_{n=1}^{\infty}\frac{z^n}{n!}$ (c) $\sum_{n=1}^{\infty}n!z^n$

(a) $R = \lim_{n\to\infty}\left|\frac{c_n}{c_{n+1}}\right| = \lim_{n\to\infty}\frac{n^2}{(n+1)^2} = 1$

(b) $R = \lim_{n\to\infty}\left|\frac{c_n}{c_{n+1}}\right| = \lim_{n\to\infty}\frac{1/n!}{1/(n+1)!} = \lim_{n\to\infty}(n+1) = +\infty$

(c) $R = \lim_{n\to\infty}\left|\frac{c_n}{c_{n+1}}\right| = \lim_{n\to\infty}\frac{n!}{(n+1)!} = \lim_{n\to\infty}\frac{1}{n+1} = 0$

**Theorem 4.2.2(Abel)** For a given power series (4.2.1), then:

(1) if $|z-z_0|<R$, the series converges absolutely;

(2) if $|z-z_0|>R$, the terms of the series become unbounded and so the series diverges;

(3) if $0\leqslant\rho<R$, then the series converges uniformly on $|z-z_0|\leqslant\rho$.

**Theorem 4.2.3** Let

$$f(z) = \sum_{n=0}^{\infty}c_n(z-z_0)^n \quad (4.2.3)$$

be a power series whose radius of convergence is $R$. Then $f(z)$ is analytic on the open disc $0<|z-z_0|<R$.

**Proof** We may assume that $f(z) = \sum_{n=0}^{\infty}c_n z^n$. We have to show that $f(z)$ has a power series expansion at an arbitrary point $z_1$ of the disc, so $|z_1|<r$. Let $s>0$ be such that $|z_1|+s<r$. We shall see that $f(z)$ can be represented by a convergent power series at $z_1$, converging absolutely on a disc of radius $s$ centered at $z_1$.

We write

$$z = z_1 + (z-z_1)$$

so that
$$z^n = [z_1 + (z-z_1)]^n$$
Then
$$f(z) = \sum_{n=0}^{\infty} c_n \left[ \sum_{k=0}^{n} \binom{n}{k} z_1^{n-k} (z-z_1)^k \right]$$

If $|z-z_1| < s$ then $|z_1| + |z-z_1| < r$, and hence the series
$$\sum_{n=0}^{\infty} |a_n|(|z_1| + |z-z_1|)^n = \sum_{n=0}^{\infty} |c_k| \sum_{k=0}^{n} \binom{n}{k} z_1^{n-k}(z-z_1)^k$$

Converges. Then we can interchange the order of summation, to get
$$f(z) = \sum_{k=0}^{\infty} \left[ \sum_{n=k}^{\infty} c_k \binom{n}{k} z_1^{n-k} \right] (z-z_1)^k$$

Which converges absolutely also, as was to be shown.

**Corollary 4.2.4** A power series (4.2.1) represents a continuous function $f(z)$ at each point inside its circle of convergence $|z-z_0| = R$.

**Theorem 4.2.5** Let $C$ denote any contour interior to the circle of convergence of the power series (4.2.1), and let $g(z)$ be any function that is continuous on $C$. The series formed by multiplying each term of the power series by $g(z)$ can be integrated term by term over $C$; that is,
$$\int_C g(z)S(z)dz = \sum_{n=0}^{\infty} c_n \int_C g(z)(z-z_0)^n dz \qquad (4.2.4)$$

**Theorem 4.2.6** The power series (4.2.1) can be differentiated term by term. That is, at each point $z$ interior to the circle of convergence of that series,
$$f'(z) = \sum_{n=1}^{\infty} nc_n(z-z_0)^{n-1} \qquad (4.2.5)$$

**Theorem 4.2.7** If the power series (4.2.1) has radius $R$, then the formal derived series
$$\sum_{n=1}^{\infty} nc_n(z-z_0)^{n-1} \qquad (4.2.6)$$
and the formally integrated series
$$\sum_{n=0}^{\infty} \frac{c_n}{n+1}(z-z_0)^{n+1} \qquad (4.2.7)$$
has the same radius $R$.

**Example 4** For any $R>0$, the series
$$\sum_{n=1}^{\infty} \frac{z^n}{n!}$$
converges absolutely and uniformly for $|z|\leq R$.
**Proof** Indeed, let
$$c_n = \frac{R^n}{n!}$$
Then
$$\frac{c_{n+1}}{c_n} = \frac{R^{n+1}}{(n+1)!} \bigg/ \frac{R^n}{n!} = \frac{R}{n+1}$$
Taken $n \geq 2R$. Then
$$\frac{c_{n+1}}{c_n} = \frac{R^{n+1}}{(n+1)!} \bigg/ \frac{R^n}{n!} = \frac{R}{n+1} \leq \frac{1}{2}$$
Hence for all $n$ sufficiently large, we have
$$c_{n+1} \leq \frac{1}{2} c_n$$
Therefore there exists constant $M$ and positive integer $n_0$ such that
$$c_n \leq \frac{M}{2^{n-n_0}}$$
By the convergence of geometric series, we complete the proof.

**Example 5 (The Binomial Series)** Let $\alpha$ be any nonzero complex number. Define the binomial coefficients as
$$\binom{\alpha}{n} = \frac{\alpha(\alpha-1)\cdots(\alpha-n+1)}{n!}$$
and the binomial series
$$(1+z)^\alpha = \sum_{n=0}^{\infty} \binom{\alpha}{n} z^n$$
By convention
$$\binom{\alpha}{0} = 1$$
To prove that the radius of convergence of the binomial series is 1 if $\alpha$ is not equal to an positive integer.
**Proof** Under the stated assumption, none of the coefficients $c_n$ are zero, and we have

$$\lim_{n\to\infty}\left|\frac{c_{n+1}}{c_n}\right| = \lim_{n\to\infty}\left|1-\frac{\alpha}{n}\right| = 1$$

## Exercises

1. Determine the radius of convergence for the following power series

   (a) $\displaystyle\sum_{n=1}^{\infty}\frac{z^n}{n}$  (b) $\displaystyle\sum_{n=1}^{\infty}\frac{nz^n}{2^n}$  (c) $\displaystyle\sum_{n=1}^{\infty}n^n z^n$  (d) $\displaystyle\sum_{n=1}^{\infty}n^p z^n$

   (e) $\displaystyle\sum_{n=0}^{\infty}\frac{1}{n!}z^n$  (f) $\displaystyle\sum_{n=0}^{\infty}n! z^n$  (g) $\displaystyle\sum_{n=0}^{\infty}z^{n!}$  (h) $\displaystyle\sum_{n=1}^{\infty}\frac{(n!)^3}{(3n)!}z^n$

2. Let $\displaystyle\sum_{n=0}^{\infty}c_n(z-z_0)^n$ have radius of convergence $R>0$. Show that the following series have the same radius of convergence

   (a) $\displaystyle\sum_{n=1}^{\infty}nc_n(z-z_0)^{n-1}$  (b) $\displaystyle\sum_{n=0}^{\infty}\frac{c_n}{n+1}z^{n+1}$  (c) $\displaystyle\sum_{n=1}^{\infty}n^2 c_n z^n$

3. If $\displaystyle\sum_{n=0}^{\infty}c_n z^n$ has radius of convergence $R$, what is the radius of convergence of $\displaystyle\sum_{n=0}^{\infty}c_n z^{2n}$ ? of $\displaystyle\sum_{n=0}^{\infty}c_n^2 z^n$ ?

4. If $\displaystyle\sum_{n=0}^{\infty}a_n z^n$ and $\displaystyle\sum_{n=0}^{\infty}b_n z^n$ have radius of convergence $R_1$ and $R_2$, show that the radius of convergence of $\displaystyle\sum_{n=0}^{\infty}a_n b_n z^n$ is at least $R_1 R_2$.

5. Show that the power series for $\log(1+z)$ converges absolutely for $|z|<1$.

6. Given an example of a power series whose radius of convergence is 1, and such that the corresponding function is continuous on the close unit disc.

7. Let $a$, $b$ be two complex numbers, and assume that $b$ is not equal to any negative integer. Show that the radius of convergence of the series
$$\sum_{n=1}^{\infty}\frac{a(a+1)(a+2)\cdots(a+n)}{b(b+1)(b+2)\cdots(b+n)}$$
is at least 1. Show that this radius can be $\infty$ in some cases.

8. Let $\{c_n\}$ be a decreasing sequence of positive numbers approaching 0. Prove that the power series $\displaystyle\sum_{n=1}^{\infty}c_n z^n$ is uniformly convergent on the domain of $z$ such that

$|z| \leqslant 1$ and $|z-1| \geqslant \delta$

where $\delta > 0$.

9. Let $a$, $b$ be two complex numbers with $|a| \leqslant |b|$. Let

$$f(z) = \sum_{n=1}^{\infty} (3a^n - 5b^n) z^n$$

Determine the radius of convergence of $f(z)$.

10. Let $\{c_n\}$ be the sequence of real numbers defined by the conditions:

$$c_0 = 1, \quad c_1 = 2, \quad \text{and} \quad c_n = c_{n-1} + c_{n-2} \quad \text{for} \quad n \geqslant 2.$$

Determine the radius of convergence of the power series

$$\sum_{n=1}^{\infty} c_n z^n$$

## 3  Taylor Series

We turn now to Taylor's theorem. We will show now that every function can be developed in a convergent Taylor series.

**Theorem 4.3.1(Taylor)**  Suppose that a function $f(z)$ is analytic throughout a disk $|z - z_0| < R_0$, centered at $z_0$ and with radius $R_0$. Then $f(z)$ has the power series representation.

$$f(z) = \sum_{n=0}^{\infty} c_n (z - z_0)^n \quad (|z - z_0| < R_0) \tag{4.3.1}$$

where

$$c_n = \frac{f^{(n)}(z_0)}{n!} \quad (n = 1, 2, \cdots) \tag{4.3.2}$$

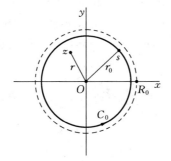

Fig. 31

With the agreement that
$$f^{(0)}(z_0) = f(z_0) \quad \text{and} \quad 0! = 1$$
series (4.3.1) can be written
$$f(z) = f(z_0) + \frac{f'(z_0)}{1!}(z-z_0) + \frac{f''(z_0)}{2!}(z-z_0)^2 + \cdots \quad (|z-z_0| < R_0)$$
(4.3.3)

(4.3.1) or (4.3.3) is called the Taylor expansion of $f(z)$ about the point $z_0$. It is the familiar Taylor series from calculus, adapted to functions of a complex variable.

When $z_0 = 0$, series (4.3.1) becomes
$$f(z) = \sum_{n=0}^{\infty} \frac{f^{(n)}(0)}{n!} z^n \quad (|z| < R_0) \tag{4.3.4}$$

(4.3.4) is called a Maclaurin series.

We now consider the uniqueness of Taylor series representations.

**Theorem 4.3.2** If a series
$$\sum_{n=0}^{\infty} c_n (z-z_0)^n \tag{4.3.5}$$
converges to $f(z)$ at all points interior to some circle $|z-z_0| = R$, then it is the Taylor series expansion for $f(z)$ in powers of $z-z_0$.

**Remark** From the statement about the radius of convergence in Theorem 4.3.1 we now see that if $R$ is the radius of convergence of a power series, then its analytic function does not extend to a disc of radius $> R$; otherwise the given power series would have a larger radius of convergence, and would represent this analytic function on the bigger disc.

We also conclude from the above remark and the theorem that if a function is entire, then its power series converges for all $z \in C$, in other words the radius of convergence is $\infty$.

**Theorem 4.3.3** Let $f(z) = \sum_{n=0}^{\infty} c_n (z-z_0)^n$ have radius of convergence $R > 0$. Then:

(1) For each $k \geq 1$ the series
$$\sum_{n=k}^{\infty} n(n-1) \cdots (n-k+1) c_n (z-z_0)^{n-k} \tag{4.3.6}$$

has radius of convergence $R$;

(2) The function $f(z)$ is infinitely differentiable on $|z-z_0|<R$ and, furthermore, $f^{(k)}(z)$ is given by the series (4.3.6) for all $k \geq 1$ and $|z-z_0|<R_0$.

**Theorem 4.3.4** If $f(z)$ is an entire function then $f(z)$ has a power series expansion

$$f(z) = \sum_{n=0}^{\infty} c_n z$$

with infinite radius of convergence.

**Example 6** To find the Maclaurin series for the function $f(z)=e^z$.

Since the function $f(z)=e^z$ is entire, it has a Maclaurin series representation which is valid for all $z$. Here $f^{(n)}(z)=e^z$; and $f^{(n)}(0)=1$, it follows that

$$e^z = \sum_{n=0}^{\infty} \frac{z^n}{n!} \quad (|z|<\infty) \tag{4.3.7}$$

**Example 7** To find the Maclaurin series for the function $f(z)=\sin z$.

Since the function $f(z)=\sin z$ is entire, it has a Maclaurin series representation which is valid for all $z$. We refer to expansion (4.3.7) and write

$$\sin z = \frac{1}{2i}\left[\sum_{n=0}^{\infty} \frac{(iz)^n}{n!} - \sum_{n=0}^{\infty} \frac{(-iz)^n}{n!}\right] = \frac{1}{2i} \sum_{n=0}^{\infty} [1-(-1)^n] \frac{i^n z^n}{n!}$$

$$= \sum_{n=0}^{\infty} (-1)^n \frac{z^{2n+1}}{(2n+1)!} \quad (|z|<\infty) \tag{4.3.8}$$

By the Theorem 4.2.7 that term by term differentiation will be used here, we differentiate each side of equation (4.3.8) and write

$$\cos z = \sum_{n=0}^{\infty} \frac{(-1)^n}{(2n+1)!} \frac{d}{dz} z^{2n+1} = \sum_{n=0}^{\infty} (-1)^n \frac{2n+1}{(2n+1)!} z^{2n}$$

That is,

$$\cos z = \sum_{n=0}^{\infty} (-1)^n \frac{z^{2n}}{(2n)!} \quad (|z|<\infty) \tag{4.3.9}$$

**Example 8** To find the Maclaurin series for the function $f(z)=\frac{1}{1-z}$.

The function $f(z)=\frac{1}{1-z}$ fails to be analytic at $z=1$, and

$$f^{(n)}(z) = \frac{n!}{(1-z)^{n+1}} \quad (n=0, 1, 2, \cdots)$$

In particular, $f^{(n)}(0) = n!$. So

$$\frac{1}{1-z} = \sum_{n=0}^{\infty} z^n \quad (|z|<1) \tag{4.3.10}$$

If we substitute $-z$ for $z$ in equation (4.3.10), we see that

$$\frac{1}{1+z} = \sum_{n=0}^{\infty} (-1)^n z^n \quad (|z|<1) \tag{4.3.11}$$

On the other hand, if we replace the variable $z$ in equation (4.3.10) by $1-z$, we have the Taylor series representation

$$\frac{1}{z} = \sum_{n=0}^{\infty} (-1)^n (z-1)^n \quad (|z-1|<1) \tag{4.3.12}$$

**Example 9** To expand the function

$$f(z) = \frac{1+2z^2}{z^3+z^5}$$

into a series involving powers of $z$.

We know from expansion (4.3.10) that

$$\frac{1}{1+z^2} = 1 - z^2 + z^4 - z^6 + z^8 - \cdots \quad (|z|<1)$$

Hence

$$f(z) = \frac{1+2z^2}{z^3+z^5} = \frac{1}{z^3} \cdot \frac{2(1+z^2)-1}{1+z^2} = \frac{1}{z^3}\left(2 - \frac{1}{1+z^2}\right)$$

$$= \frac{1}{z^3}(2 - 1 + z^2 - z^4 + z^6 - z^8 - \cdots)$$

$$= \frac{1}{z^3} + \frac{1}{z} - z + z^3 - z^5 + \cdots \quad 0<|z|<1$$

We call such terms as $1/z^3$ and $1/z$ negative powers of $z$.

**Example 10** Let

$$\cos z = 1 - \frac{z^2}{2!} + \frac{z^4}{4!} - \cdots$$

be the formal power series whose coefficients are the same as for the Taylor expansion of the ordinary cosine function in elementary calculus. To write down the first few terms of its (formal) inverse,

$$\frac{1}{\cos z}$$

up to term of order 4.

We have
$$\frac{1}{\cos z} = \frac{1}{1 - \frac{z^2}{2!} + \frac{z^4}{4!} - \cdots} = 1 + \left(\frac{z^2}{2!} - \frac{z^4}{4!} + \cdots\right) + \left(\frac{z^2}{2!} - \frac{z^4}{4!} + \cdots\right)^2 + \cdots$$

$$= 1 + \frac{z^2}{2!} - \frac{z^4}{4!} + \frac{z^4}{(2!)^2} + \text{higher terms}$$

This gives us the coefficients of $\frac{1}{\cos z}$ up to order 4.

**Example 11** Find the terms of order $\leqslant 3$ in the power series for the function
$$\frac{1}{\sin z}$$

By expansion (4.3.8), we have
$$\sin z = \sum_{n=0}^{\infty} (-1)^n \frac{z^{2n+1}}{(2n+1)!} = z\left(1 - \frac{z^2}{3!} + \frac{z^4}{5!} - \cdots\right) \quad (|z| < \infty)$$

Hence
$$\frac{1}{\sin z} = \frac{1}{z} \cdot \frac{1}{1 - \frac{z^2}{3!} + \frac{z^4}{5!} - \cdots}$$

$$= \frac{1}{z}\left(1 + \frac{z^2}{3!} - \frac{z^4}{5!} + \frac{z^4}{(3!)^2} + \text{higher terms}\right)$$

$$= \frac{1}{z} + \frac{1}{3!}z + \left(\frac{1}{(3!)^2} - \frac{1}{5!}\right)z^3 + \text{higher terms}$$

**Example 12** Find the Maclaurin series expansion of the function
$$f(z) = \frac{e^z}{1-z}$$

We know from expansion (4.3.7) and (4.3.10) that
$$f(z) = \frac{e^z}{1-z} = \sum_{n=0}^{\infty} \frac{z^n}{n!} \sum_{n=0}^{\infty} z^n = \left(1 + z + \frac{z^2}{2!} + \cdots\right)(1 + z + z^2 + \cdots)$$

$$= 1 + 2z + \frac{5}{2}z^2 + \cdots \quad |z| < 1$$

## Exercises

1. Expand the functions below in powers series $z$. What is the radius of convergence?

   (a) $\dfrac{1}{az+b}$      (b) $\displaystyle\int_0^z \frac{\sin z}{z} dz$      (c) $\displaystyle\int_0^z e^{z^2} dz$

(d) $\sin^2 z$    (e) $\dfrac{1}{(1-z)^2}$

2. Develop the functions below in powers of $z-1$. What is the radius of convergence?

   (a) $\sin z$    (b) $\dfrac{z-1}{z+1}$    (c) $\dfrac{z}{z^2-2z+5}$

   (d) $\dfrac{2z+3}{z+1}$    (e) $\sqrt{z}\,(\sqrt[3]{1}=\dfrac{-1+\sqrt{3}\,i}{2})$

3. Find the Maclaurin series expansion of the function
$$f(z)=\dfrac{z}{z^4+9}=\dfrac{z}{9}\cdot\dfrac{1}{1+(z^4/9)}$$

4. Expand $\cos z$, $\sin z$ in a power series about $\dfrac{\pi}{2}$.

5. Give the power series expansion of $\log z$ about $z=i$ and find its radius of convergence.

6. Give the power series expansion of $\sqrt{z}$ about $z=1$ and find its radius of convergence.

7. Give the terms of order $\leqslant 3$ in the power series $e^z\,\mathrm{Ln}(1+z)$, where $\mathrm{Ln}(1+z)|_{z=0}=0$.

8. Expand $\dfrac{1}{(1-z)^m}$, $m$ a positive integer, in powers of $z$ in the domain of $f(z)$.

9. For what value of $z$ is
$$\sum_{n=0}^{\infty}\left(\dfrac{z}{1+z}\right)^n$$
   Convergent?

10. For what value of $z$ is
$$\sum_{n=0}^{\infty}\dfrac{z^n}{1+z^{2n}}$$
   Convergent?

11. Show that the series
$$\sum_{n=1}^{\infty}\dfrac{z^{n-1}}{(1-z^n)(1-z^{n+1})}$$
Converges to $\dfrac{1}{(1-z)^2}$ for $|z|<1$ and to $\dfrac{1}{z(1-z)^2}$ for $|z|>1$. Prove

that the convergence is uniform for $|z|\leq c<1$ in the first case, and $|z|\geq b>1$ in the second.

12. Give the terms of order $\leq 3$ in the power series:

   (a) $e^z \sin z$    (b) $(\sin z)(\cos z)$    (c) $\dfrac{e^z-1}{z}$    (d) $\dfrac{e^z-\cos z}{z}$

   (e) $\dfrac{\cos z}{\sin z}$    (f) $\dfrac{\sin z}{\cos z}$    (g) $\dfrac{e^z}{\sin z}$

13. Find the terms of order $\leq 3$ in the power series expansion of the function
$$f(z)=\frac{z^2}{z-2} \text{ at } z=1$$

14. Find the terms of order $\leq 3$ in the power series expansion of the function
$$f(z)=\frac{z-2}{(z+3)(z+2)} \text{ at } z=1$$

15. Let
$$f(z)=\sum_{n=0}^{\infty} c_n z^n$$

Define
$$f(-z)=\sum_{n=0}^{\infty} c_n (-z)^n = \sum_{n=0}^{\infty} (-1)^n c_n z^n$$

We define $f(z)$ to be even if $c_n=0$ for $n$ odd. We define $f(z)$ to be odd if $c_n=0$ for $n$ even. Verify that $f(z)$ is even if and only if $f(-z)=f(z)$ and $f(z)$ is odd if and only if $f(-z)=-f(z)$.

16. Let
$$f(z)=\sum_{n=0}^{\infty} \frac{z^{2n}}{(2n)!}$$

Prove that $f''(z)=f(z)$.

17. Let
$$f(z)=\sum_{n=0}^{\infty} \frac{z^{2n}}{(n!)^2}$$

Prove that $z^2 f''(z)+z f'(z)=4z^2 f(z)$.

18. Let
$$f(z)=z-\frac{z^3}{3}+\frac{z^5}{5}-\frac{z^7}{7}+\cdots$$

Show that $f'(z)=\dfrac{1}{z^2+1}$.

19. Let
$$f(z) = \sum_{n=0}^{\infty} \frac{(-1)^n}{(n!)^2} \left(\frac{z}{2}\right)^{2n}$$
Prove that $z^2 f''(z) + z f'(z) + z^2 f(z) = 0$.

20. Prove that if $f(z)$ is analytic at $z_0$ and $f(z_0) = f'(z_0) = \cdots = f^{(m)}(z_0) = 0$, then the function $g(z)$ defined by the equations
$$g(z) = \begin{cases} \dfrac{f(z)}{(z-z_0)^{m+1}}, & \text{when } z \ne z_0 \\ \dfrac{f^{(m+1)}(z_0)}{(m+1)!}, & \text{when } z = z_0 \end{cases}$$
is analytic at $z_0$.

21. Suppose that a function $f(z)$ has a power series representation
$$f(z) = \sum_{n=0}^{\infty} a_n (z-z_0)^n$$
inside some circle $|z-z_0| = R$. To show that
$$f^{(n)}(z) = \sum_{k=0}^{\infty} \frac{(n+k)!}{k!} a_{n+k} (z-z_0)^k \quad (n = 0, 1, 2, \cdots)$$
when $|z-z_0| < R$. Then, by setting $z = z_0$, show that the coefficients $a_n$ ($n=0, 1, 2, \cdots$) are the coefficients in the Taylor series for $f(z)$ about $z_0$.

22. Show that the radius of convergence of the power series
$$\sum_{n=1}^{\infty} \frac{(-1)^n}{n} z^{n(n+1)}$$
is 1, and discuss convergence for $z=1$, $-1$ and $z=i$.

23. (a) Let $\sum_{n=0}^{\infty} c_n (z-z_0)^n$ have radius of convergence 1 and suppose that $\sum_{n=0}^{\infty} c_n$ converges to S. Prove that $\lim_{r \to 1^-} \sum_{n=0}^{\infty} c_n r^n = S$.

(b) To prove that
$$\log 2 = 1 - \frac{1}{2} + \frac{1}{3} - \cdots$$

24. Show that the series
$$f(z) = \sum_{n=1}^{\infty} \left(\frac{z+i}{z-i}\right)^n$$

defined an analytic function a disc of radius 1 centered at $-i$.

## 4  Laurent Series

If a function $f(z)$ fails to be analytic at point $z_0$, we cannot apply Taylor's theorem at that point. It is often to find a series representation for $f(z)$ involving both positive and negative powers of $z-z_0$. We now present the Laurent's theorem.

**Theorem 4.4.1(Laurent)**  Suppose that a function $f(z)$ is analytic throughout an annular domain $R_1 < |z-z_0| < R_2$, centered at $z_0$, and let $C$ denote any positively oriented simple closed contour around $z_0$, and lying in that domain. Then, at each point in the domain, $f(z)$ has the series representation

$$f(z) = \sum_{n=-\infty}^{\infty} c_n (z-z_0)^n \quad (R_1 < |z-z_0| < R_2) \quad (4.4.1)$$

where

$$c_n = \frac{1}{2\pi i} \int_C \frac{f(z)dz}{(z-z_0)^{n+1}} \quad (n=0, \pm 1, \pm 2 \cdots) \quad (4.4.2)$$

The forms (4.4.1) is called a Laurent series.

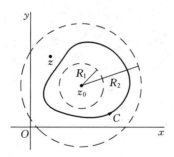

Fig. 32

**Theorem 4.4.2**  If a series

$$\sum_{n=-\infty}^{\infty} c_n (z-z_0)^n = \sum_{n=0}^{\infty} a_n (z-z_0)^n + \sum_{n=1}^{\infty} \frac{b_n}{(z-z_0)^n} \quad (4.4.3)$$

converges to $f(z)$ at all points in some annular domain about $z_0$, then it is the Laurent series expansion for $f(z)$ in powers of $z-z_0$ for that domain.

**Example 13** To find the Laurent series for the function
$$f(z)=\frac{1}{z(z-1)}$$
where $0<|z|<1$ and $|z|>1$.

We write $f(z)$ in partial fractions
$$f(z)=\frac{1}{z(z-1)}=\frac{1}{z-1}-\frac{1}{z}$$
Then for one term we get the geometric series,
$$\frac{1}{z-1}=-\frac{1}{1-z}=-(1+z+z^2+\cdots)$$
Whence
$$f(z)=-\frac{1}{z}-1-z^2-\cdots$$
On the other hand, we write
$$\frac{1}{z-1}=\frac{1}{z}\left[\frac{1}{1-\frac{1}{z}}\right]=\frac{1}{z}\left(1+\frac{1}{z}+\frac{1}{z^2}+\cdots\right)$$
Hence
$$f(z)=\frac{1}{z^2}+\frac{1}{z^3}+\frac{1}{z^4}+\cdots$$

**Example 14** To find the Laurent series for the function
$$f(z)=\frac{1}{z-1}-\frac{1}{z-2}$$
where $|z|<1$, $1<|z|<2$, and $2<|z|<\infty$.

The function
$$f(z)=\frac{1}{z-1}-\frac{1}{z-2}$$
has the two singular points $z=1$ and $z=2$, is analytic in the domains
$$|z|<1, \quad 1<|z|<2, \quad \text{and} \quad 2<|z|<\infty$$
When $|z|<1$, the representation is a Maclaurin series. To find it, we write
$$f(z)=-\frac{1}{1-z}+\frac{1}{2}\cdot\frac{1}{1-(z/2)}$$
and observe that, since $|z|<1$ and $\left|\frac{z}{2}\right|<1$ in $|z|<1$,
$$f(z)=-\sum_{n=0}^{\infty}z^n+\sum_{n=0}^{\infty}\frac{z^n}{2^{n+1}}=\sum_{n=0}^{\infty}(2^{-n-1}-1)z^n \quad (|z|<1)$$

When $1<|z|<2$, the representation is a Laurent series. We write
$$f(z)=\frac{1}{z}\cdot\frac{1}{1-(1/z)}+\frac{1}{2}\cdot\frac{1}{1-(z/2)}$$
Since $\left|\frac{1}{z}\right|<1$ and $\left|\frac{z}{2}\right|<1$ when $1<|z|<2$, it follows that
$$f(z)=\sum_{n=0}^{\infty}\frac{1}{z^{n+1}}+\sum_{n=0}^{\infty}\frac{z^n}{2^{n+1}}\quad(1<|z|<2)$$

The representation of $f(z)$ in $2<|z|<\infty$ is also a Laurent series. If we write
$$f(z)=\frac{1}{z}\cdot\frac{1}{1-(1/z)}-\frac{1}{z}\cdot\frac{1}{1-(2/z)}$$
and observe that $\left|\frac{1}{z}\right|<1$ and $\left|\frac{2}{z}\right|<1$ when $2<|z|<\infty$, we find that
$$f(z)=\sum_{n=0}^{\infty}\frac{1}{z^{n+1}}-\sum_{n=0}^{\infty}\frac{2^n}{z^{n+1}}=\sum_{n=0}^{\infty}\frac{1-2^n}{z^{n+1}}=\sum_{n=1}^{\infty}\frac{1-2^n}{z^{n+1}}$$

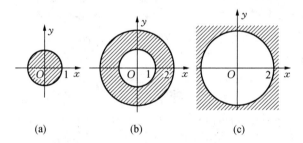

(a)　　　　(b)　　　　(c)

Fig. 33

## Exercises

1. Give the Laurent expansion for the following the functions
   (a) $\dfrac{z+1}{z^2(z-1)}$, $0<|z|<1$; $1<|z|<+\infty$;
   (b) $\dfrac{z^2-2z+5}{(z-2)(z^2+1)}$, $1<|z|<2$;
   (c) $\dfrac{e^z}{z(z^2+1)}$, $0<|z|<1$, Find the terms of order $\leqslant 2$.

2. Find the Laurent series that represents the function
$$f(z)=z^2\sin\left(\frac{1}{z^2}\right)$$

in the domain $0 < |z| < \infty$.

3. Derive the Laurent series representation
$$\frac{e^z}{(z+1)^2} = \frac{1}{e}\left[\sum_{n=0}^{\infty}\frac{(z+1)^n}{(n+2)!} + \frac{1}{z+1} + \frac{1}{(z+1)^2}\right] \quad (0 < |z+1| < \infty)$$

4. Find a representation for the function
$$f(z) = \frac{1}{1+z} = \frac{1}{z} \cdot \frac{1}{1+(1/z)}$$
in negative powers of $z$ that is valid when $1 < |z| < \infty$.

5. Give two Laurent series expansions in powers of $z$ for the function
$$f(z) = \frac{1}{z^2(1-z)}$$
and specify the regions in which those expansions are valid.

6. Represent the function
$$f(z) = \frac{z+1}{z-1}$$
   (a) by its Maclaurin series, and state where the representation is valid;
   (b) by it Laurent series in the domain $1 < |z| < \infty$.

7. Show that when $0 < |z-1| < 2$,
$$\frac{z}{(z-1)(z-3)} = -3\sum_{n=0}^{\infty}\frac{(z-1)^n}{2^{n+2}} - \frac{1}{2(z-1)}$$

8. Write two Laurent series in powers of $z$ that represent the function
$$f(z) = \frac{1}{z(1+z^2)}$$
in certain domains and specify those domains.

9. Expand the functions below in Laurent series at the point
   (a) $\dfrac{1}{(z^2+1)^2}$, $z=i$    (b) $z^2 e^{\frac{1}{z}}$, $z=0$, $z=\infty$
   (c) $e^{\frac{1}{1-z}}$, $z=1$, $z=\infty$

10. Give the Laurent series for $\dfrac{z+1}{z-1}$ in the region
    (a) $|z| < 1$    (b) $|z| > 1$

11. Give the Laurent expansions for the following functions:
    (a) $\dfrac{z}{z+2}$ for $|z| > 2$    (b) $\sin\dfrac{1}{z}$ for $z \neq 0$

(c) $\cos \dfrac{1}{z}$ for $z \neq 0$ \qquad (d) $\dfrac{1}{z-3}$ for $|z|>3$

12. Find the Laurent series for $\dfrac{1}{z^2(1-z)}$ in the region

    (a) $0<|z|<1$ \qquad (b) $|z|>1$

13. Find the Laurent expansion of
$$f(z)=\dfrac{1}{(z-1)^2(z+1)^2}$$
for $1<|z|<2$.

14. Obtain the first four terms of the Laurent series expansion of
$$f(z)=\dfrac{e^z}{z(z^2+1)}$$
valid for $0<|z|<1$.

15. Expand the function
$$f(z)=\dfrac{z}{1+z^3}$$

    (a) in a series of positive powers of $z$, and
    (b) in a series of negative powers of $z$.

16. Find the power series expansion of
$$f(z)=\dfrac{1}{1+z^2}$$
around the point $z=1$, and find the radius of convergence of this series.

17. Prove that the Laurent development is unique.

# 5 Zeros of an Analytic Functions and Uniquely Determined Analytic Functions

## 5.1 Zeros of Analytic Functions

If $p(z)$ and $q(z)$ are two polynomials then it is well known that $p(z)=s(z)q(z)+r(z)$ where $s(z)$ and $r(z)$ are also polynomials and the degree of $r(z)$ is less than the degree of $q(z)$.

**Definition 4.5.1** If $f(z)$ and all derivatives $f^{(n)}(z)$ $(n=1, 2, \cdots)$ exist at $z_0$, we can write

$$f(z)=(z-z_0)^m g(z)$$

for any $n$. Assume that $f(z)$ is not identically zero. Then, if $f(z_0)=0$, there exists a first derivative $f^{(m)}(z_0)$ which is different from zero. We say that the point $z_0$ is a zero of order $m$.

**Theorem 4.5.1** Suppose that a function $f(z)$ is analytic at a point $z_0$. Then $f(z)$ has a zero of order $m$ at $z_0$ if and only if there is a function $g(z)$, which is analytic and nonzero at $z_0$, such that

$$f(z)=(z-z_0)^m g(z) \qquad (4.5.1)$$

**Proof** First, we assume that expression (4.5.1) holds, since $g(z)$ is analytic at $z_0$, it has a Taylor series representation

$$g(z)=g(z_0)+\frac{g'(z_0)}{1!}(z-z_0)+\frac{g''(z_0)}{2!}(z-z_0)^2+\cdots$$

in some neighborhood $|z-z_0|<\varepsilon$ of $z_0$. Expression (4.5.1) thus takes the from

$$f(z)=g(z_0)(z-z_0)^m+\frac{g'(z_0)}{1!}(z-z_0)^{m+1}+\frac{g''(z_0)}{2!}(z-z_0)^{m+2}+\cdots$$

when $|z-z_0|<\varepsilon$. It follows that

$$f(z_0)=f'(z_0)=f''(z_0)=\cdots=f^{(m-1)}(z_0)=0 \qquad (4.5.2)$$

and that

$$f^{(m)}(z_0)=m!\ g(z_0)\neq 0 \qquad (4.5.3)$$

Hence $z_0$ is a zero of order $m$ of $f(z)$.

Conversely, if we assume that $f(z)$ has a zero of order $m$ at $z_0$, its analyticity at $z_0$ and the fact that conditions (4.5.2) hold tell us that, in some neighborhood $|z-z_0|<\varepsilon$, there is a Taylor series

$$f(z) = \sum_{n=m}^{\infty} \frac{f^{(n)}(z_0)}{n!}(z-z_0)^n$$

$$= (z-z_0)^m\left[\frac{f^m(z_0)}{m!}+\frac{f^{(m+1)}(z_0)}{(m+1)!}(z-z_0)+\frac{f^{(m+2)}(z_0)}{(m+2)!}(z-z_0)^2+\cdots\right]$$

Consequently, $f(z)$ has the form (4.5.1), where

$$g(z)=\frac{f^{(m)}(z_0)}{m!}+\frac{f^{(m+1)}(z_0)}{(m+1)!}(z-z_0)+\frac{f^{(m+2)}(z_0)}{(m+2)!}(z-z_0)^2+\cdots(|z-z_0|<\varepsilon)$$

The convergence of this last series when $|z-z_0|<\varepsilon$ ensures that $g(z)$ is analytic in that neighborhood and, in particular, at $z_0$. Moreover,

$$g(z_0) = \frac{f^{(m)}(z_0)}{m!} \neq 0$$

This completes the proof of the theorem.

**Remark** There are no zeros of infinite order.

**Example 15** The entire function $f(z) = z(e^z - 1)$ has a zero of order $m=2$ at the point $z_0 = 0$.

**Proof** since
$$f(0) = f'(0) = 0 \text{ and } f''(0) = 2 \neq 0$$

The function $g(z)$ in expression (4.5.1), defined by the equations
$$g(z) = \begin{cases} \dfrac{e^z - 1}{z}, & \text{when } z \neq 0 \\ 1, & \text{when } z = 0 \end{cases}$$

It is analytic at $z=0$ and is entire.

**Example 16** The entire function $f(z) = z - \sin z$ has a zero of order $m=3$ at the point $z_0 = 0$.

**Proof** We know from expansion (4.3.8) that
$$f(z) = z - \sin z = z - \sum_{n=0}^{\infty} (-1)^n \frac{z^{2n+1}}{(2n+1)!}$$
$$= z - \left(z - \frac{z^3}{3!} + \frac{z^5}{5!} - \cdots\right)$$
$$= z^3 \left(\frac{1}{3!} - \frac{z^2}{5!} + \cdots\right)$$

We can define the function $g(z)$ as
$$g(z) = \begin{cases} \dfrac{z - \sin z}{z^3}, & \text{when } z \neq 0 \\ \dfrac{1}{3!}, & \text{when } z = 0 \end{cases}$$

Which is analytic at $z=0$ and is entire.

The Theorem 4.5.2, Theorem 4.5.3, Theorem 4.5.4 and Theorem 4.5.5 are from Reference [2].

The next theorem tells us that the zeros of an analytic function are isolated.

**Theorem 4.5.2** Given a function $f(z)$ and a point $z_0$, suppose that

(1) $f(z)$ is analytic at $z_0$;

(2) $f(z_0)=0$ but $f(z)$ is not identically equal to zero in any neighborhood of $z_0$.

Then $f(z) \neq 0$ throughout some deleted neighborhood $0 < |z-z_0| < \varepsilon$ of $z_0$.

**Proof** Let $f(z)$ be as stated and observe that not all of the derivatives of $f(z)$ at $z_0$ are zero. For, if they were, all of the coefficients in the Taylor series for $f(z)$ about $z_0$ wound be zero; and that would mean that $f(z)$ is identically equal to zero in some neighborhood of $z_0$. So it is clear from the definition of zeros of order $m$ at the beginning of this section that $f(z)$ must have a zero of some order $m$ at $z_0$. According to Theorem 1, then,

$$f(z)=(z-z_0)^m g(z) \qquad (4.5.4)$$

where $g(z)$ is analytic and nonzero at $z_0$.

Now $g$ is continuous, in addition to being nonzero, at $z_0$ because it is analytic there. Hence there is some neighborhood $|z-z_0| < \varepsilon$ in which equation (4.5.4) holds and in which $g(z) \neq 0$. Consequently, $f(z) \neq 0$ in the deleted neighborhood $0 < |z-z_0| < \varepsilon$; and the proof is complete.

The next theorem here concerns functions with zeros that are not all isolated.

**Theorem 4.5.3** Suppose that $f$ is analytic throughout a neighborhood $N_0$ of $z_0$ and $f(z_0)=0$. If $f(z)=0$ at each point $z$ of a domain or line segment containing $z_0$, then $f(z) \equiv 0$ in $N_0$; that is, $f(z)$ is identically equal to zero throughout $N_0$.

**Proof** Since under the stated conditions, $f(z) \equiv 0$ in some neighborhood $N$ of $z_0$. For, otherwise, there would be a deleted neighborhood of $z_0$ throughout which $f(z) \neq 0$, according to Theorem 4.5.2 above; and that would be inconsistent with the condition that $f(z)=0$ everywhere in a domain or on a line segment containing $z_0$. Since $f(z) \equiv 0$ in the neighborhood $N$, then, it follows that all of the coefficients

$$a_n = \frac{f^{(n)}(z_0)}{n!} \quad (n=0, 1, 2, \cdots)$$

in the Taylor series for $f(z)$ about $z_0$ must be zero. Thus $f(z) \equiv 0$ in the neighborhood $N_0$, since Taylor series also represents $f(z)$ in $N_0$. This completes the proof.

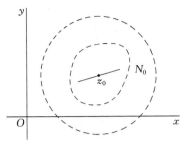

**Fig. 34**

## 5.2 Uniquely Determined Analytic Functions

**Theorem 4.5.4** A function that is analytic on a domain $D$ is uniquely determined over $D$ by its values in a domain, or along a line segment, contained in $D$.

**Theorem 4.5.5** Suppose that

(1) $f(z)$ is analytic on a domain $D$;

(2) $f(z)=0$ on a domain or a line segment contained in $D$.

Then $f(z) \equiv 0$ in $D$.

**Proof** Let $f(z)$ be as stated in its hypothesis and let $z_0$ be any point of the subdomain or line segment at each point of which $f(z)=0$. Since $D$ is a connected open set, there is a polygonal line $L$, consisting of a finite number of line segments joined end to end and lying entirely in $D$, that extends from $z_0$ to any other point $P$ in $D$. We let $d$ be the shortest distance from points on $L$ to the boundary of $D$, unless $D$ is the entire plane; in that case, $d$ may be positive number. We then form a finite sequence of points

$$z_0, z_1, z_2, \cdots, z_{n-1}, z_n$$

along $L$, where the point $z_n$ coincides with $P$ and where each points is sufficiently close to the adjacent ones that

$$|z_k - z_{k-1}| < d \quad (k=1, 2, \cdots, n)$$

Finally, we construct a finite sequence of neighborhoods

$$N_0, N_1, N_2, \cdots, N_{n-1}, N_n$$

where each neighborhood $N_k$ is centered at $z_k$ and has radius $d$. Note that these neighborhoods are all contained in $D$ and that the center $z_k$ of any

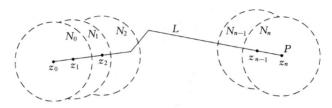

Fig. 35

neighborhood $N_k$ ($k = 1, 2, \cdots, n$) lies in the preceding neighborhood $N_{k-1}$.

At this point, since $f(z)$ is analytic in the domain $N_0$ and since $f(z) = 0$ in a domain or on a line segment containing $z_0$, then $f(z) \equiv 0$ in $N_0$. But the point $z_1$ lies in the domain $N_0$. Hence a second application of the same theorem reveals that $f(z) \equiv 0$ in $N_n$. Since $N_n$ is centered at the point $P$ and since $P$ was arbitrarily selected in $D$, we may conclude that $f(z) \equiv 0$ in $D$. This completes the proof of the lemma.

**Theorem 4.5.6** Suppose now that two functions $f(z)$ and $g(z)$ are analytic in the same domain $D$ and that $f(z) = g(z)$ at each point $z$ of some domain or line segment contained in $D$. Then $f(z) \equiv 0$ in $D$.

This follows by applying the preceding theorem to the analytic function $h(z) = f(z) - g(z)$

**Proof** The difference

$$h(z) = f(z) - g(z)$$

is also analytic in $D$, and $h(z) = 0$ throughout the subdomain or along the line segment. According to the above lemma, then $h(z) = 0$ throughout $D$; that is, $f(z) = g(z)$ at each point $z$ in $D$.

**Theorem 4.5.7** If $f(z)$ is analytic on a domain $D$ and $f(z)$ is not identically zero then for each $z_0$ in $D$ with $f(z_0) = 0$ there is an integer $n \geq 1$ and an analytic function $g(z)$ such that $g(z_0) \neq 0$ and $f(z) = (z - z_0)^n g(z)$ for all $z$ in $D$. That is, each zero of $f(z)$ has finite multiplicity.

**Proof** Let $n \geq 1$ be the largest integer such that $f^{(n-1)}(z_0) = 0$ and define

$$g(z) = \frac{f(z)}{(z - z_0)^n}$$

for $z\neq z_0$ and $g(z_0)=\dfrac{f^{(n)}(z_0)}{n!}$. Then $g(z)$ is clearly analytic in in a neighborhood of $z_0$. This completes the proof.

## 5.3 Maximum Modulus Principle

**Theorem 4.5.8 (Mean Value Theorem)** The value of an analytic function at the center of a circle is equal to the arithmetic mean of its values on the circle, subject to the condition that the closed disk $|z-z_0|\leqslant r$ is contained in the region of analyticity.

**Proof** We write
$$z=z_0+re^{i\theta}, \quad dz=ire^{i\theta}$$
Then by Cauchy's integral formula,
$$f(z_0)=\frac{1}{2\pi i}\int_{|z-z_0|=r}\frac{f(z)}{z-z_0}dz$$
$$=\frac{1}{2\pi i}\int_0^{2\pi}\frac{f(z_0+re^{i\theta})}{re^{i\theta}}ire^{i\theta}d\theta$$
$$=\frac{1}{2\pi}\int_0^{2\pi}f(z_0+re^{i\theta})d\theta$$

**Theorem 4.5.9 (Maximum Modulus Principle)** If a function $f(z)$ is analytic and not constant in a given domain $D$, then $|f(z)|$ has no maximum value in $D$. That is, there is no point $z_0$ in the domain such that $|f(z)|\leqslant|f(z_0)|$ for all points $z$ in it.

**Proof** To prove this, we assume that $f(z)$ satisfies the stated conditions and let $z_1$ be any point other than $z_0$ in the given neighborhood. We then let $\rho$ be the distance between $z_1$ and $z_0$. If $C_\rho$ denotes the positively oriented circle $|z-z_0|=\rho$, centered at $z_0$ and passing through $z_1$, the Theorem 4.5.7 tells us that
$$|f(z_0)|\leqslant\frac{1}{2\pi}\int_0^{2\pi}|f(z_0+\rho e^{i\theta})|d\theta \tag{4.5.5}$$
On the other hand, since
$$|f(z_0+\rho e^{i\theta})|\leqslant|f(z_0)| \quad (0\leqslant\theta\leqslant 2\pi) \tag{4.5.6}$$
we find that
$$\int_0^{2\pi}|f(z_0+\rho e^{i\theta})|d\theta\leqslant\int_0^{2\pi}|f(z_0)|d\theta=2\pi|f(z_0)|$$

Thus
$$|f(z_0)| \geq \frac{1}{2\pi}\int_0^{2\pi} |f(z_0+\rho e^{i\theta})|\, d\theta \tag{4.5.7}$$

It is now evident from inequalities (4.5.6) and (4.5.7) that
$$|f(z_0)| = \frac{1}{2\pi}\int_0^{2\pi} |f(z_0+\rho e^{i\theta})|\, d\theta \tag{4.5.8}$$

therefore,
$$|f(z_0+\rho e^{i\theta})| = |f(z_0)| \quad (0 \leq \theta \leq 2\pi) \tag{4.5.9}$$

This shows that $|f(z)| = |f(z_0)|$ for all points $z$ on the circle $|z-z_0|=\rho$.

Finally, since $z$ is any point in the deleted neighborhood $0 < |z-z_0| < \varepsilon$, we see that the equation $|f(z)| = |f(z_0)|$ is, in fact, satisfied by all points $z$ lying on any circle $|z-z_0|=\rho$, where $0<\rho<\varepsilon$. Consequently, we get $|f(z)| = |f(z_0)|$ everywhere in the neighborhood $|z-z_0|<\varepsilon$. we that when the modulus of an analytic function is constant in a domain, the function itself is constant there. Thus $f(z)=f(z_0)$ for each point $z$ in the neighborhood, and the proof of the Theorem is complete.

**Corollary 4.5.10** Suppose that a function $f(z)$ is continuous on a closed bounded region $R$ and that it is analytic and not constant in the interior of $R$. Then the maximum value of $|f(z)|$ on $R$, which is always reached, occurs somewhere on the boundary of $R$ and never in the interior.

**Example 17** Suppose the function $f(z)=u(x, y)+iv(x, y)$ is continuous on a closed bounded region $R$. Then the component function $u(x, y)$ has a maximum value in $R$.

**Proof** Let $g(z)=e^{f(z)}$, then the function $g(z)$ is continuous in $R$ and analytic and not constant in the interior. Since
$$|g(z)| = e^{u(x,y)}$$
is continuous in $R$, so $g|(z)|$ must achieve maximum value on the boundary. It follows that the maximum value of $u(x, y)$ also occurs on the boundary.

## Exercises

1. Find the order of the zeros of the functions at the point $z=0$
   (a) $z^2(e^{z^2}-1)$
   (b) $6\sin z^3 + z^3(z^6-6)$

## Chapter IV Series

2. Let $f(z)$ be an entire function such that $|f(z)| \leq A|z|$ for all $z$, where $A$ is a fixed positive number. Show that $f(z)=a_1 z$, where $a_1$ is a complex constant.

3. Suppose that $f(z)$ is entire and that the harmonic function $u(x, y)=\operatorname{Re}[f(z)]$ has an upper bound $u_0$; that is, $u(x, y) \leq u_0$ for all points $(x, y)$ in the $xy$ plane. Show that $u(x, y)$ must be constant through the plane.

4. Let a function $f(z)$ be continuous in a closed bounded region $R$, and let it be analytic and not constant throughout the interior of $R$. Assuming that $f(z) \neq 0$ anywhere in $R$, prove that $|f(z)|$ has a minimum value $m$ on $R$ which occurs on the boundary of $R$ and never in the interior. This result is named Minimum Modulus Principle.

5. Let the function $f(z)=u(x, y)+iv(x, y)$ be continuous on a closed bounded region $R$, and suppose that it is analytic and not constant in the interior of $R$. Show that the component function $v(x, y)$ has maximum and minimum values in $R$ which are reached on the boundary of $R$ and never in the interior, where it is harmonic.

6. Let $f(z)$ be an entire function and suppose there is a constant $M$, $R>0$ and an integer $n \geq 1$ such that $|f(z)| \leq M|z|^n$ for $|z|>R$. Show that $f(z)$ is a polynomial of degree $\leq n$.

7. Let $D$ be a region and suppose that $f(z)$ is analytic and $z_0 \in D$ such that $|f(z_0)| \leq |f(z)|$ for all $z$ in $D$. Show that either $f(z_0)=0$ or $f(z)$ is constant.

8. Give an elementary proof of the Maximum Modulus Theorem for polynomials.

9. Let $D$ be a region and let $f(z)$ and $g(z)$ be analytic functions on $D$ such that $f(z)=\overline{g(z)}$ for all $z$ in $D$. Show that either $f(z)=0$ or $g(z)=0$.

10. Show that if $f(z)$ and $g(z)$ are analytic functions on a region $D$ such that $\overline{f(z)}g(z)$ is analytic then either $f(z)$ is constant or $g(z) \equiv 0$.

11. If $f(z)$ is a non-constant analytic function on a bounded open set $D$ and is continuous on $\overline{D}$, then either $f(z)$ has a zero in $D$ or $|f(z)|$ assumes its minimum value on $\partial D$.

12. Let $D$ be a bounded region and suppose $f(z)$ is continuous on $\overline{D}$ and ana-

lytic on $D$. Show that if there is a constant $c \geq 0$ such that $|f(z)| = c$ for all $z$ on the boundary of $D$ then either $f(z)$ is a constant function or $f(z)$ has a zero in $D$.

13. (a) Let $f(z)$ be entire and non-constant. For any positive real number $c$ show that the closure of $\{z: |f(z)| < c\}$ is the set $\{z: |f(z)| \leq c\}$.

    (b) Let $p(z)$ be a polynomial and show that each component of $\{z: |p(z)| < c\}$ contains a zero of $p(z)$.

    (c) If $p(z)$ is a polynomial and $c > 0$ show that $\{z: |p(z)| = c\}$ is the union of a finite number of closed paths. Discuss the behavior of these paths as $c \to \infty$.

14. Let $f(z)$ be analytic on $|z| \leq R$ with $|f(z)| \leq M$ for $|z| \leq R$ and $|f(0)| = a > 0$. Show that the number of zeros of $f(z)$ in $|z| \leq R$ is less than or equal to $\frac{1}{\log 2} \log\left(\frac{M}{a}\right)$.

15. Suppose that both $f(z)$ and $g(z)$ are analytic on $|z| \leq R$ with $|f(z)| = |g(z)|$ for $|z| = R$.
    
    Show that if neither $f(z)$ nor $g(z)$ vanishes in $|z| \leq R$ then there is a constant $\lambda$, $|\lambda| = 1$, such that $f(z) = \lambda g(z)$.

16. Let $f(z)$ be analytic in the disk $|z| \leq R$ and for $0 < r \leq R$ define $A(r) = \max\{\operatorname{Re} f(z): |z| = r\}$. Show that unless $f(z)$ is a constant, $A(r)$ is a strictly increasing function of $r$.

17. Let $U$ be a connected open set, and let $D$ be an open disc whose closure is contained in $U$. Let $f(z)$ be analytic on $U$ and not constant. Assume that the absolute value $|f(z)|$ is constant on the boundary of $D$. Prove that $f(z)$ has at least one zero in $D$.

# 6 The Three Types of Isolated Singular Points at a Finite Point

**Definition 4.6.1** Let $z_0$ be a complex number and if a function $f(z)$ fails to be analytic at a point $z_0$ but is analytic at some point in every neighborhood of $z_0$, then $z_0$ is called a singular point, or singularity of $f(z)$.

**Definition 4.6.2** A singular point $z_0$ is said to be isolated if, in addition, there is a deleted neighborhood $0<|z-z_0|<\varepsilon$ of $z_0$ throughout which $f(z)$ is analytic.

**Definition 4.6.3** Let $D$ be a domain and let $U$ be the open set obtained by removing $z_0$ from $D$. A function $f(z)$ which is analytic on $U$ is said to have an isolated singularity at $z_0$.

**Example 18** The functions $\frac{\sin z}{z}, \frac{1}{z}$ and $e^{\frac{1}{z}}$ all have isolated singularities at $z=0$.

**Example 19** The function
$$\frac{z+1}{z(z^2+1)}$$
has the three isolated singular points $z=0$ and $z=\pm i$.

**Example 20** The function
$$\frac{1}{\sin\frac{\pi}{z}}$$
has the singular point $z=0$ and $z=1/n (n=\pm 1, \pm 2, \cdots)$. Each singular point $z=1/n (n=\pm 1, \pm 2, \cdots)$ is isolated. The singular point $z=0$ is not isolated because every deleted $\varepsilon$ neighborhood of the origin contains other singular points of the function.

**Definition 4.6.4** If $f(z)$ has an isolated singular point $z_0$, then $f(z)$ can be represented by a Laurent series
$$f(z) = \sum_{n=-\infty}^{\infty} c_n(z-z_0)^n \qquad (4.6.1)$$
in a punctured disk $0<|z-z_0|<R$. The portion
$$\frac{c_{-1}}{z-z_0} + \frac{c_{-2}}{(z-z_0)^2} + \cdots + \frac{c_{-n}}{(z-z_0)^n} + \cdots$$
of the series, involving negative powers of $z-z_0$, is called the principal part of $f(z)$ at $z_0$.

**Definition 4.6.5** If expansion (4.6.1) takes the from
$$f(z) = \sum_{n=0}^{\infty} c_n(z-z_0)^n + \frac{c_{-1}}{z-z_0} + \frac{c_{-2}}{(z-z_0)^2} + \cdots + \frac{c_{-m}}{(z-z_0)^m}$$
$$(4.6.2)$$
where $c_{-m} \neq 0$ and $0<|z-z_0|<R_2$. In this case, the isolated singular point $z_0$ is called a pole of order $m$. A pole of order $m=1$ is usually referred

to as a simple pole.

**Theorem 4.6.1** A function $f(z)$ has a pole of order $m$ at $z_0$ if and only if
$$f(z) = \frac{\phi(z)}{(z-z_0)^m}$$
where $\phi(z)$ is analytic and nonzero at $z_0$.

The proof is immediate and is left to the students.

**Theorem 4.6.2** If $z_0$ is a pole of a function $f(z)$, then
$$\lim_{z \to z_0} f(z) = \infty \qquad (4.6.3)$$

**Proof** To verify limit (4.6.3), we assume that $f(z)$ has a pole of order $m$ at $z_0$. It tells us that
$$f(z) = \frac{\phi(z)}{(z-z_0)^m}$$
where $\phi(z)$ is analytic and nonzero at $z_0$. Since
$$\lim_{z \to z_0} \frac{1}{f(z)} = \lim_{z \to z_0} \frac{(z-z_0)^m}{\phi(z)} = \frac{\lim_{z \to z_0}(z-z_0)^m}{\lim_{z \to z_0}\phi(z)} = \frac{0}{\phi(z_0)} = 0$$
then, limit (4.6.3) holds, according to the theorem in Chapter 1 regarding limits that involve the point at infinity. This completes the proof.

**Example 21** The function
$$\frac{z^2-2z+3}{z-2} = \frac{z(z-2)+3}{z-2} = z + \frac{3}{z-2} = 2 + (z-2) + \frac{3}{z-2} \quad (0 < |z-2| < \infty)$$
has a simple pole ($m=1$) at $z_0 = 2$.

**Example 22** The function $\frac{1}{z}$ has a simple pole at the origin.

**Example 23** The function $\frac{1}{\sin z}$ has a simple pole at the origin.

Because
$$\sin z = \sum_{n=0}^{\infty} (-1)^n \frac{z^{2n+1}}{(2n+1)!} = z + \text{higher terms}$$
and
$$\frac{1}{\sin z} = \frac{1}{z} + \text{higher terms}$$

**Example 24** The function
$$\frac{\sh z}{z^4} = \frac{1}{z^4}\left(z + \frac{z^3}{3!} + \frac{z^5}{5!} + \frac{z^7}{7!} + \cdots\right) = \frac{1}{z^3} + \frac{1}{3!} \cdot \frac{1}{z} + \frac{z}{5!} + \frac{z^3}{7!} + \cdots \quad (0 < |z| < \infty)$$

has a pole of order $m=3$ at $z_0=0$.

**Definition 4.6.6** If expansion (4.6.1) takes the form
$$f(z) = \sum_{n=0}^{\infty} c_n(z-z_0)^n \quad (0 < |z-z_0| < R) \tag{4.6.4}$$
the point $z_0$ is known as a removable singular point.

**Remark** The residue at a removable singular point is always zero. If we define $f(z_0) = a_0$, expansion (4.6.4) becomes valid throughout the entire disk $|z-z_0| < R$. Since a power series always represents an analytic function interior to its circle of convergence, it follows that $f(z)$ is analytic at $z_0$ when it is assigned the value $a_0$ there. The singularity at $z_0$ is, therefore, removed.

**Theorem 4.6.3** If $f(z)$ has an isolated singularity at $z_0$ then the point $z_0$ is a removable singularity if
$$\lim_{z \to z_0}(z-z_0)f(z) = 0$$

**Proof** Suppose $f(z)$ is analytic in $\{z: 0 < |z-z_0| < R\}$, and define $g(z) = (z-z_0)f(z)$ for $z \neq z_0$ and $g(z_0) = 0$. Suppose $\lim_{z \to z_0}(z-z_0)f(z) = 0$; then $g(z)$ is clearly a continuous function. If we can show that $g(z)$ is analytic then it follows that $a$ is a removable singularity. In fact, if $g(z)$ is analytic we have $g(z) = (z-z_0)h(z)$ for some analytic function defined on $\{z: |z-z_0| < R\}$ because $g(z_0) = 0$. But then $h(z)$ and $f(z)$ must agree for $0 < |z-z_0| < R$, so that $z_0$, by definition, a removable singularity.

**Theorem 4.6.4** If $z_0$ is a removable singular point of a function $f(z)$, then $f(z)$ is analytic and bounded in some deleted neighborhood $0 < |z-z_0| < \varepsilon$ of $z_0$.

**Proof** Let $z_0 = 0$. We know that $f(z)$ has a Laurent expansion
$$f(z) = \sum_{n=-\infty}^{\infty} c_n z^n$$
for $0 < |z| < R$. We have to show $c_n = 0$ if $n < 0$. Let $n = -m$ with $m > 0$. We have
$$c_{-m} = \frac{1}{2\pi i} \int_{C_R} f(s) s^{m-1} ds$$
Since $f(z)$ is assumed bounded near zero, it follows that the right-hand side tends to zero as $R$ tends to zero, whence $c_{-m} = 0$, as was to be shown. The

proof is completed.

**Theorem 4.6.5 (Riemann)** Suppose that a function $f(z)$ is analytic and bounded in some deleted neighborhood $0 < |z - z_0| < \varepsilon$ of a point $z_0$. If $f(z)$ is not analytic at $z_0$, then it has a removable singularity there.

**Proof** To prove this, we assume that $f(z)$ is not analytic at $z_0$. As a consequence, the point $z_0$ must be an isolated singularity of $f(z)$; and $f(z)$ is represented by a Laurent series

$$f(z) = \sum_{n=0}^{\infty} a_n (z - z_0)^n + \sum_{n=1}^{\infty} \frac{b_n}{(z - z_0)^n} \qquad (4.6.5)$$

throughout the deleted neighborhood $0 < |z - z_0| < \varepsilon$. If $C$ denotes a positively oriented circle $|z - z_0| = \rho$, where $\rho < \varepsilon$, the coefficients $b_n$ in expansion (4.6.5) can be written

$$b_n = \frac{1}{2\pi i} \int_C \frac{f(z)}{(z - z_0)^{-n+1}} dz \quad (n = 0, 1, 2, \cdots) \qquad (4.6.6)$$

Now the boundedness on $f(z)$ tells us that there is a positive constant $M$ such that $|f(z)| \leq M$ whenever $0 < |z - z_0| < \varepsilon$. Hence it follows from expression (4.6.6) that

$$|b_n| \leq \frac{1}{2\pi} \frac{M}{\rho^{-n+1}} 2\pi\rho = M\rho^n \quad (n = 1, 2, \cdots)$$

Since the coefficients $b_n$ are constants and since $\rho$ can be chosen arbitrarily small, we may conclude that $b_n = 0$ $(n = 1, 2, \cdots)$ in the Laurent series (4.6.5). This tells us that $z_0$ is a removable singularity of $f(z)$, and the proof of the lemma is complete.

**Example 25** The point $z_0 = 0$ is a removable singular point of the function

$$f(z) = \frac{1 - \cos z}{z^2} = \frac{1}{z^2} \left[ 1 - \left( 1 - \frac{z^2}{2!} + \frac{z^4}{4!} - \frac{z^6}{6!} + \cdots \right) \right]$$

$$= \frac{1}{2!} - \frac{z^2}{4!} + \frac{z^4}{6!} - \cdots \quad (0 < |z| < \infty)$$

When we define $f(0) = 1/2$, $f(z)$ becomes an entire function.

**Definition 4.6.7** When an infinite number of the coefficients $c_{-n}$ in the principal part are nonzero, $z_0$ is said to be an essential singular point of $f(z)$.

**Theorem 4.6.6 (Casorati-Weierstrass)** Suppose that $z_0$ is an essential singularity of a function $f(z)$, and let $w_0$ be any complex number. Then, for any positive number $\varepsilon$, the inequality

$$|f(z)-w_0|<\varepsilon \qquad (4.6.7)$$

is satisfied at some point $z$ in each deleted neighborhood $0<|z-z_0|<\delta$ of $z_0$. In other words, the values of $f(z)$ come arbitrarily close to any complex number.

**Proof** Suppose the theorem is false. Since $z_0$ is an isolated singularity of $f(z)$, there is a deleted neighborhood $0<|z-z_0|<\delta$ throughout which $f(z)$ is analytic; and we assume that condition (4.6.7) is not satisfied for any point $z$ there. Thus $|f(z)-w_0|\geq\varepsilon$ when $0<|z-z_0|<\delta$; and so the function

$$g(z)=\frac{1}{f(z)-w_0}\ (0<|z-z_0|<\delta) \qquad (4.6.8)$$

is bounded and analytic in its domain of definition. Hence, According to the Theorem 4.6.4, $z_0$ is a removable singularity of $g(z)$; and $g(z)$ may be extended to a analytic function. It then follows that $\dfrac{1}{g(z)}$ has at most a pole at $z_0$, which means that $f(z)-w_0$ has at most a pole, contradicting the hypothesis that $f(z)$ has an essential singularity (infinitely many terms of negative order in its Laurent series). This proves the theorem.

$$\text{isolated singular points}\begin{cases}\text{removable singular points}\\ \text{poles}\\ \text{essential singular points}\end{cases}$$

**Example 26** The function $e^{\frac{1}{z}}$ has an essential singular point at $z=0$ because its Laurent series is

$$e^{\frac{1}{z}}=\sum_{n=0}^{\infty}\frac{1}{n!}\cdot\frac{1}{z^n}=1+\frac{1}{1!}\cdot\frac{1}{z}+\frac{1}{2!}\cdot\frac{1}{z^2}+\cdots(0<|z|<\infty)$$

**Theorem 4.6.7** Let $z=z_0$ be an isolated singularity of $f(z)$ and let $f(z)=\sum_{n=-\infty}^{\infty}c_n(z-z_0)^n$ be its Laurent expansion in $0<|z-z_0|<R$. Then:

(1) $z=z_0$ is a removable singularity if $c_n=0$ for $n\leq-1$;

(2) $z=z_0$ is a pole of order if $c_{-m}\neq 0$ and $c_n=0$ for $n\leq-(m+1)$;

(3) $z=z_0$ is an essential singularity if $c_n\neq 0$ for infinitely many negative integers $n$.

## Exercises

1. In each case, determine whether that point is a pole, a removable point, or an essential singular point

   (a) $\dfrac{z-1}{z(z^2+4)^2}$     (b) $\dfrac{1}{\sin z + \cos z}$     (c) $\dfrac{1-e^z}{1+e^z}$

   (d) $\dfrac{1}{(z^2+i)^3}$     (e) $\tan^2 z$     (f) $\cos \dfrac{1}{z+i}$

   (g) $\dfrac{1-\cos z}{z^2}$     (h) $\dfrac{1}{e^z - 1}$

2. In each case, determine whether that point is a pole, a removable point, or an essential singular point;

   (a) $\dfrac{1}{e^z - 1} - \dfrac{1}{z}$     (b) $e^{z - \frac{1}{z}}$     (c) $\sin \dfrac{1}{z} + \dfrac{1}{z^2}$

   (d) $e^z \cos \dfrac{1}{z}$     (e) $\dfrac{e^{\frac{1}{z-1}}}{e^z - 1}$

3. Determine the set of poles, and find their orders.

   (a) $\displaystyle\sum_{n=0}^{\infty} \dfrac{(-1)^n}{n!(z^2+n^2)}$     (b) $\displaystyle\sum_{n=0}^{\infty} \dfrac{1}{z^2+n^2}$     (c) $\dfrac{1}{(z+n)^2}$

   (d) $\displaystyle\sum_{n=1}^{\infty} \dfrac{\sin nz}{n!(z^2+n^2)}$     (e) $\dfrac{1}{z} + \displaystyle\sum_{\substack{n=-\infty \\ n \neq 0}}^{\infty} \left[\dfrac{1}{z-n} + \dfrac{1}{n}\right]$

4. Show that the function
$$f(z) = \sum_{n=1}^{\infty} \dfrac{z^2}{n^2 z^2 + 8}$$
is defined and continuous for the real values of $z$. Determine the region of the complex plane in which this function is analytic. Determine its poles.

5. If $f(z)$ and $g(z)$ have the algebraic orders $h$ and $k$ at $z=a$, show that $f(z)g(z)$ has the order $h+k$, $\dfrac{f(z)}{g(z)}$ the order $h-k$, and $f(z)+g(z)$ an order which does not exceeded $\max(h, k)$.

6. Show that an isolated singularity of $f(z)$ cannot be a pole of $e^{f(z)}$.

7. Each of the following functions $f(z)$ has an isolated singularity at $z=0$. Determine its nature; if it is a removable singularity define $f(0)$ so that $f(z)$ is analytic at $z=0$; if it is a pole find the singular part; if it is an

essential singularity determine $f(\{z: 0<|z|<\delta\})$ for arbitrarily small values of $\delta$.

(a) $f(z) = \dfrac{\sin z}{z}$  (b) $f(z) = \dfrac{\cos z}{z}$  (c) $f(z) = \dfrac{\cos z - 1}{z}$

(d) $f(z) = e^{\frac{1}{z}}$  (e) $f(z) = \dfrac{\log(z+1)}{z^2}$  (f) $f(z) = \dfrac{\cos z^{-1}}{z^{-1}}$

(g) $f(z) = \dfrac{z^2}{z(z-1)}$  (h) $f(z) = \dfrac{1}{1-e^z}$  (i) $f(z) = z\sin\dfrac{1}{z}$

(j) $f(z) = z^n \sin\dfrac{1}{z}$

8. Give the partial fraction expansion of
$$f(z) = \dfrac{z^2+1}{(z^2+z+1)(z-1)^2}$$

9. If $f(z)$ is analytic function on a domain $D$ and except for poles. Show that the poles of $f(z)$ cannot have a limit point in $D$.

10. In each case, write the principal part of the function at its isolated singular point and determine whether that point is a pole, a removable point, or an essential singular point;

(a) $ze^{\frac{1}{z}}$  (b) $\dfrac{z^2}{1+z}$  (c) $\dfrac{\sin z}{z}$  (d) $\dfrac{\cos z}{z}$  (e) $\dfrac{1}{(2-z)^3}$

# 7 The Three Types of Isolated Singular Points at a Infinite Point

**Definition 4.7.1** For each small positive number $\varepsilon$. We thus call the set $D = \{z: |z| > 1/\varepsilon\}$ an $\varepsilon$ - neighborhood, or neighborhood, of $\infty$.

**Definition 4.7.2** Suppose the function $f(z)$ which is analytic on $D = \{z: |z| > 1/\varepsilon\}$, then we say $f(z)$ have an isolated singularity at $\infty$.

Let $S$ be the union of $C$ and a single other point denoted by $\infty$, and called infinity. Suppose $f(z)$ be a function on $S$. Let $t = \dfrac{1}{z}$, and define
$$g(t) = f\left(\dfrac{1}{t}\right)$$
for $t \neq 0, \infty$. We say that $f(z)$ has an isolated singularity at infinity if $g(t)$ has an isolated singularity at $t = 0$. The function $f(z)$ has a removable singu-

larity, a pole, or an essential singularity at infinity if $g(t)$ has, respectively, a removable singularity, a pole, or an essential singularity at $t=0$. If $f(z)$ has a pole at $\infty$ then the order of the pole is the order of the pole of $g(t)$ at $t=0$. By this way, we can extend the conclusions about the three types of isolated singular points at a finite point to infinity.

**Theorem 4.7.1**

(1) An entire function has a removable singularity at infinity if it is a constant.

(2) An entire function has a pole at infinity of order $m$ if it is a polynomial of degree $m$.

(3) Characterize those rational functions which have a removable singularity at infinity.

(4) Characterize those rational functions which have a pole of order $m$ at infinity.

## Exercises

1. Show that a function which is analytic in the whole plane and has a nonessential singularity at $\infty$ reduces to a polynomial.
2. Show that the functions $e^z$, $\sin z$, $\cos z$ have essential singularities at $\infty$.
3. Show that any function which meromorphic in the extended plane is rational.

# Chapter V

## Calculus of Residues

In this chapter, a specific number, called a residue, which each of those points contributes to the value of the integral, is introduced. We develop here the theory of residues; and then apply them to evaluate certain types of definite and improper integrals over the real line which cannot be evaluated by the methods of calculus.

## 1 Residues

### 1.1 Residues

**Definition 5.1.1** If $f(z)$ have an isolated singularity at $z=z_0$ and let

$$f(z) = \sum_{n=-\infty}^{\infty} c_n(z-z_0)^n$$

be its Laurent expansion at $z=z_0$. Then the residue of $f(z)$ at $z=z_0$ is the coefficient $c_{-1}$. Denote this by

$$\operatorname*{Res}_{z=z_0} f(z) = c_{-1} \tag{5.1.1}$$

The coefficients $c_{-1}$ have certain integral representations. In particular,

$$\int_C f(z)\,dz = 2\pi i c_{-1} \tag{5.1.2}$$

where $C$ is any positively oriented simple closed contour around $z_0$ and lying in the punctured disk $0 < |z-z_0| < R$.

Equation (5.1.2) provides a powerful method for evaluating certain integrals around simple closed contours.

**Example 1** Find the residue of

$$f(z) = \frac{\sin z}{z^2} \text{ at } z=0$$

We have

$$f(z) = \frac{\sin z}{z^2} = \frac{1}{z^2}\left(z - \frac{z^3}{3!} + \cdots\right) = \frac{1}{z} + \text{higher terms}$$

Hence the desired residue is 1.

**Example 2** Find the residue of
$$f(z) = \frac{z^2}{(z+1)(z-1)^2} \quad \text{at } z=1$$

We can write
$$f(z) = \frac{[(z-1)+1]^2}{[(z-1)+2](z-1)^2} = \frac{1}{2(z-1)^2} + \frac{\frac{3}{4}}{z-1} + \cdots$$

Thus
$$\operatorname*{Res}_{z=1} f(z) = \frac{3}{4}$$

**Example 3** To evaluate the integral
$$\int_C \frac{dz}{z(z-2)^4}$$
where $C$ is the positively oriented circle $|z-2|=1$.

Since the integrand is analytic everywhere in the finite plane except at the points $z=0$ and $z=2$, only $z=2$ is interior to $C$. We write
$$\frac{1}{z(z-2)^4} = \frac{1}{(z-2)^4} \cdot \frac{1}{2+(z-2)}$$
$$= \frac{1}{2(z-2)^4} \cdot \frac{1}{1-\left(-\frac{z-2}{2}\right)}$$
$$= \sum_{n=0}^{\infty} \frac{(-1)^n}{2^{n+1}} (z-2)^{n-4} \quad (0 < |z-2| < 1)$$

Consequently,
$$\int_C \frac{dz}{z(z-2)^4} = 2\pi i\left(-\frac{1}{16}\right) = -\frac{\pi i}{8}$$

**Example 4** Let $C$ be a circle centered at 1, of radius 1. To find the integral
$$\int_C \frac{z^2}{(z+1)(z-1)^2} dz$$

Since the function $f(z)$ has two singularities at $z=1$ and $z=-1$, only $z=1$ is interior to $C$. Hence
$$\int_C \frac{z^2}{(z+1)(z-1)^2} dz = 2\pi i \operatorname*{Res}_{z=1} f(z) = 2\pi i \cdot \frac{3}{4} = \frac{3}{2} \pi i$$

**Example 5** Find the integral
$$\int_C e^{\frac{1}{z^2}} dz$$
where $C$ is the unit circle $|z|=1$.

Since the integrand is analytic everywhere in the finite plane except at the point $z=0$, and $z=0$ is interior to $C$. We can write the Laurent series expansion
$$e^{\frac{1}{z^2}} = 1 + \frac{1}{1!} \cdot \frac{1}{z^2} + \frac{1}{2!} \cdot \frac{1}{z^4} + \frac{1}{3!} \cdot \frac{1}{z^6} + \cdots (0<|z|<\infty)$$
So
$$\int_C e^{\frac{1}{z^2}} dz = 0$$

## 1.2 Cauchy's Residue Theorem

Cauchy's integral formula can be considered as a special case of the residue theorem.

**Theorem 5.1.1 (Cauchy's Residue Theorem)** Let $C$ be a simple closed contour, described in the positive sense. If a function $f(z)$ is analytic inside and on $C$ except for a finite number of singular points $z_k (k=1, 2, \cdots, n)$ inside $C$, then
$$\int_C f(z) dz = 2\pi i \sum_{k=1}^{n} \operatorname*{Res}_{z=z_k} f(z) \qquad (5.1.3)$$

**Proof** Let the point $z_k (k=1, 2, \cdots, n)$ be centers of positively oriented circles $C_k$ which are interior to $C$ and are so small that no two of them have points in common (Fig. 36). The circles $C_k$, together with the simple closed contour $C$, form the boundary of a closed region throughout which $f(z)$ is analytic and whose interior is a multiply connected domain. Hence, according to Theorem 3.2.6, we have
$$\int_C f(z) dz = \sum_{k=1}^{n} \int_{C_k} f(z) dz$$
Because
$$\int_{C_k} f(z) dz = 2\pi i \operatorname*{Res}_{z=z_k} f(z) \quad (k=1,2,\cdots,n)$$

and the proof is complete.

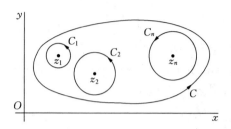

Fig. 36

**Example 6** To evaluate the integral

$$\int_C \frac{5z-2}{z(z-1)} dz$$

where $C$ is the circle $|z|=2$, described in the counterclockwise directions.

The integrand has the two isolated singularities $z=0$ and $z=1$, both of which are interior to $C$. Suppose

$$f(z) = \frac{5z-2}{z(z-1)}$$

Then, when $0<|z|<1$,

$$\frac{5z-2}{z(z-1)} = \frac{5z-2}{z} \cdot \frac{-1}{1-z} = \left(5 - \frac{2}{z}\right)(-1-z-z^2-\cdots)$$

We find that

$$\operatorname*{Res}_{z=0} f(z) = 2$$

Also, when $0<|z-1|<1$,

$$\frac{5z-2}{z(z-1)} = \frac{5(z-1)+3}{z-1} \cdot \frac{1}{1+(z-1)}$$

$$= \left(5 + \frac{3}{z-1}\right)[1-(z-1)+(z-1)^2-\cdots]$$

It is clear that

$$\operatorname*{Res}_{z=1} f(z) = 3$$

Thus

$$\int_C \frac{5z-2}{z(z-1)} dz = 2\pi i [\operatorname*{Res}_{z=0} f(z) + \operatorname*{Res}_{z=1} f(z)] = 10\pi i$$

Also, we can write the integrand as

$$\frac{5z-2}{z(z-1)} = \frac{2}{z} + \frac{3}{z-1}$$

Then

$$\int_C \frac{5z-2}{z(z-1)} dz = 2\pi i \cdot 2 + 2\pi i \cdot 3 = 10\pi i$$

## 1.3 The Calculus of Residue

(1) The Residue at Infinity

**Theorem 5.1.2** If a function $f(z)$ is analytic everywhere in the finite plane except for a finite number of singular points interior to a positively oriented simple closed contour $C$, then

$$\int_C f(z) dz = 2\pi i \operatorname*{Res}_{z=0}\left[\frac{1}{z^2} f\left(\frac{1}{z}\right)\right] \tag{5.1.4}$$

**Proof** We construct a circle $|z| = R_1$ which is large enough so that the contour $C$ is interior to it (Fig. 37). Then if $C_0$ denotes a positively oriented circle $|z| = R_0$, where $R_0 > R_1$, we know from Laurent's theorem that

$$f(z) = \sum_{n=-\infty}^{\infty} c_n z^n \quad (R_1 < |z| < \infty) \tag{5.1.5}$$

where

$$c_n = \frac{1}{2\pi i} \int_{C_0} \frac{f(z) dz}{z^{n+1}} \quad (n = 0, \pm 1, \pm 2, \cdots) \tag{5.1.6}$$

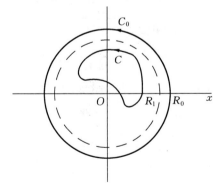

**Fig. 37**

If we replace $z$ by $1/z$ in representation (5.1.5), we see that

$$\frac{1}{z^2} f\left(\frac{1}{z}\right) = \sum_{n=-\infty}^{\infty} \frac{c_n}{z^{n+2}} = \sum_{n=-\infty}^{\infty} \frac{c_{n-2}}{z^n} \left(0 < |z| < \frac{1}{R_1}\right)$$

and hence that

$$c_{-1}=\operatorname*{Res}_{z=0}\left[\frac{1}{z^2}f\left(\frac{1}{z}\right)\right] \qquad (5.1.7)$$

Then, in view of equations (5.1.6) and (5.1.7),

$$\int_{C_0} f(z)dz = 2\pi i \operatorname*{Res}_{z=0}\left[\frac{1}{z^2}f\left(\frac{1}{z}\right)\right]$$

Finally, since $f(z)$ is analytic throughout the closed region bounded by $C$ and $C_0$, by the Theorem 3.2.5, the proof is completed.

**Example 7** Use the Theorem 5.1.2 to evaluate the example 6.

Since

$$\frac{1}{z^2}f\left(\frac{1}{z}\right)=\frac{5-2z}{z(1-z)}=\frac{5-2z}{z}\cdot\frac{1}{1-z}$$

$$=\left(\frac{5}{z}-2\right)(1+z+z^2+\cdots)$$

$$=\frac{5}{z}+3+3z+\cdots(0<|z|<1)$$

We see that

$$\operatorname*{Res}_{z=0}\frac{1}{z^2}f\left(\frac{1}{z}\right)=5$$

So

$$\int_C \frac{5z-2}{z(z-1)}dz = 2\pi i \cdot 5 = 10\pi i$$

**Example 8** To evaluate the integral

$$\int_C \frac{z^{15}}{(z^2+1)^2(z^4+2)^3}dz$$

where $C$ is the circle $|z|=4$, described in the counterclockwise directions.

Suppose

$$f(z)=\frac{z^{15}}{(z^2+1)^2(z^4+2)^3}$$

Since

$$\frac{1}{z^2}f\left(\frac{1}{z}\right)=\frac{1}{z^2}\cdot\frac{z^{-15}}{(z^{-2}+1)^2(z^{-4}+2)^3}$$

$$=\frac{1}{z}\cdot\frac{1}{(1+z^2)^2(1+2z^4)^3}$$

Chapter V  Calculus of Residues

$$= \frac{1}{z}(1-2z_2+\cdots)(1-3.2z^4+\cdots)$$

Then

$$\mathop{\mathrm{Res}}_{z=0} \frac{1}{z^2} f\left(\frac{1}{z}\right) = 1$$

So

$$\int_C \frac{z^{15}}{(z^2+1)^2(z^4+2)^3} dz = 2\pi i \mathop{\mathrm{Res}}_{z=0}\left[\frac{1}{z^2}f\left(\frac{1}{z}\right)\right] = 2\pi i$$

(2) Residues at Poles

**Theorem 5.1.3** Suppose $f(z)$ has a pole of order $m$ at $z=z_0$ and $f(z)$ can be written in the form

$$f(z) = \frac{\phi(z)}{(z-z_0)^m} \qquad (5.1.8)$$

where $\phi(z)$ is analytic and nonzero at $z_0$. Then

$$\mathop{\mathrm{Res}}_{z=z_0} f(z) = \frac{\phi^{(m-1)}(z_0)}{(m-1)!} \quad if \quad m \geqslant 1 \qquad (5.1.9)$$

**Proof**  We assume that $f(z)$ has the form (5.1.8) and since $\phi(z)$ is analytic at $z_0$, it has a Taylor series representation

$$\phi(z) = \phi(z_0) + \frac{\phi'(z_0)}{1!}(z-z_0) + \frac{\phi''(z_0)}{2!}(z-z_0)^2 + \cdots$$
$$+ \frac{\phi^{(m-1)}(z_0)}{(m-1)!}(z-z_0)^{m-1} + \sum_{n=m}^{\infty} \frac{\phi^{(n)}(z_0)}{n!}(z-z_0)^n$$

in some neighborhood $|z-z_0|<\varepsilon$ of $z_0$; and from expression (5.1.8) it follows that

$$f(z) = \frac{\phi(z_0)}{(z-z_0)^m} + \frac{\phi'(z_0)/1!}{(z-z_0)^{m-1}} + \frac{\phi''(z_0)/2!}{(z-z_0)^{m-2}} + \cdots$$
$$+ \frac{\phi^{(m-1)}(z_0)/(m-1)!}{z-z_0} + \sum_{n=m}^{\infty} \frac{\phi^{(n)}(z_0)}{n!}(z-z_0)^{n-m}$$

(5.1.10)

when $0<|z-z_0|<\varepsilon$.

Then we have

$$\mathop{\mathrm{Res}}_{z=z_0} f(z) = \frac{\phi^{(m-1)}(z_0)}{(m-1)!} \quad if \quad m \geqslant 1$$

**Example 9**  To find the residue of the function

$$f(z) = \frac{5z-2}{z(z-1)^2}$$

at its isolated singular point.

The function $f(z)$ has the isolated singular point $z=0$ and $z=1$. The point $z=0$ is a simple pole of the function $f(z)$, and can be written as

$$f(z)=\frac{\phi(z)}{z} \quad \text{where} \quad \phi(z)=\frac{5z-2}{(z-1)^2}$$

Since $\phi(z)$ is analytic at $z=0$ and $\phi(0)=-2\neq 0$.
So

$$\operatorname*{Res}_{z=0} f(z)=-2$$

The point $z=1$ is a pole of the second order, and can be written as

$$f(z)=\frac{\phi(z)}{(z-1)^2} \quad \text{where} \quad \phi(z)=\frac{5z-2}{z}$$

Since $\phi(z)$ is analytic at $z=1$ and $\phi(1)=3\neq 0$.
So

$$\operatorname*{Res}_{z=1} f(z)=\phi'(1)=2$$

**Example 10**   To find the residue of the function

$$f(z)=\frac{\operatorname{sh}z}{z^4}$$

at the point $z=0$.

It would be incorrect to write

$$f(z)=\frac{\phi(z)}{z^4} \quad \text{where} \quad \phi(z)=\operatorname{sh}z$$

and to attempt an application of formula (5.1.9) with $m=4$. For it is necessary that $\phi(z_0)\neq 0$ if that formula is to be used.

In this case, we use the Laurent series for $f(z)$. Then $z=0$ is a pole of the third order, with residue $\dfrac{1}{6}$.

**Example 11**   To find the residue of the function

$$f(z)=\frac{1}{z(e^z-1)}$$

at the point $z=0$.

Consider the function

$$f(z)=\frac{1}{z(e^z-1)}=\frac{1}{z^2\left(1+\dfrac{z}{2!}+\dfrac{z^2}{3!}+\cdots\right)}=\frac{\phi(z)}{z^2}$$

## Chapter V  Calculus of Residues

where $\phi(z) = \dfrac{1}{1 + \dfrac{z}{2!} + \dfrac{z^2}{3!} + \cdots}$

Since the functions $\phi(z)$ is analytic at the point $z_0 = 0$ and $\phi(0) = 1 \neq 0$. So $f(z)$ has a pole of order $m = 2$ at the point $z_0 = 0$. Thus

$$\operatorname*{Res}_{z=0} f(z) = \operatorname*{Res}_{z=0} \dfrac{1}{z(e^z - 1)} = \phi'(0) = -\dfrac{1}{2}$$

**Example 12**  To evaluate the integral

$$\int_C \dfrac{z \sin z}{(1 - e^z)^3} dz$$

where $C$ is the circle $|z| = 1$, described in the counterclockwise directions.

(1) Since the integrand is analytic everywhere except at the point $z = 0$ interior to $C$. We write

$$f(z) = \dfrac{z \sin z}{(1 - e^z)^3} = \dfrac{z \left( z - \dfrac{z^3}{3!} + \cdots \right)}{\left[ 1 - \left( 1 + z + \dfrac{z^2}{2!} + \cdots \right) \right]^3} = -\dfrac{1}{z} - c_1 - \cdots$$

So

$$\operatorname*{Res}_{z=0} f(z) = \operatorname*{Res}_{z=0} \dfrac{z \sin z}{(1 - e^z)^3} = -1$$

Thus

$$\int_C \dfrac{z \sin z}{(1 - e^z)^3} dz = 2\pi i \operatorname*{Res}_{z=0} f(z) = -2\pi i$$

(2) Consider the function

$$f(z) = \dfrac{z \sin z}{(1 - e^z)^3} = \dfrac{1}{z} \dfrac{\dfrac{\sin z}{z}}{\left( \dfrac{1 - e^z}{z} \right)^3} = \dfrac{\phi(z)}{z}$$

where $\phi(z) = \dfrac{\dfrac{\sin z}{z}}{\left( \dfrac{1 - e^z}{z} \right)^3}$

Since the functions $\lim\limits_{z \to 0} \phi(z) = -1$, the point $z_0 = 0$ is a removable singularity. So $f(z)$ has a simple pole at the point $z_0 = 0$. Thus

$$\operatorname*{Res}_{z=0} f(z) = \operatorname*{Res}_{z=0} \dfrac{z \sin z}{(1 - e^z)^3} = \lim_{z \to 0} \phi(z) = -1$$

So
$$\int_C \frac{z\sin z}{(1-e^z)^3}dz = 2\pi i \operatorname*{Res}_{z=0} f(z) = -2\pi i$$

**Theorem 5.1.4** Let two functions $p(z)$ and $q(z)$ be analytic at a point $z_0$. If
$$p(z_0) \neq 0, \quad q(z_0) = 0, \quad \text{and} \quad q'(z_0) \neq 0$$
then $z_0$ is a simple pole of the quotient $\dfrac{p(z)}{q(z)}$ and
$$\operatorname*{Res}_{z=z_0} \frac{p(z)}{q(z)} = \frac{p(z_0)}{q'(z_0)} \tag{5.1.11}$$

**Proof** Because the point $z_0$ is a zero of order $m=1$ of the function $q(z)$. Then
$$q(z) = (z-z_0)g(z) \tag{5.1.12}$$
where $g(z)$ is analytic and nonzero at $z_0$. Then
$$\frac{p(z)}{q(z)} = \frac{p(z)/g(z)}{z-z_0}$$
Now $\dfrac{p(z)}{g(z)}$ is analytic and nonzero at $z_0$, thus
$$\operatorname*{Res}_{z=z_0} \frac{p(z)}{q(z)} = \frac{p(z_0)}{g(z_0)} \tag{5.1.13}$$
But $g(z_0) = q'(z_0)$. The proof is complete.

**Example 13** Let the function
$$f(z) = \frac{1}{\sin z}$$
Because $q'(z) = \cos z$ and $q'(k\pi) = -1 \neq 0$. Then $q(z)$ has a simple zero at $z = k\pi$, and
$$\operatorname*{Res}_{z=k\pi} f(z) = \frac{1}{\cos k\pi} = (-1)^k$$

**Example 14** Find the residue of the function
$$f(z) = \frac{z^2}{z^2-1} \quad \text{at } z=1$$
The function $f(z)$ has a simple pole at $z=1$, and
$$\operatorname*{Res}_{z=1} f(z) = \frac{z^2}{z^2-1} = \frac{1}{2}$$
To do this, we can also write

## Chapter V  Calculus of Residues

$$f(z) = \frac{g(z)}{z-1}, \quad \text{where} \quad g(z) = \frac{z^2}{z+1}$$

Note that $g(z)$ is analytic at $z=1$, and that $g(1)=1\neq 0$. Hence

$$\operatorname*{Res}_{z=1} f(z) = g(1) = \frac{1}{2}$$

**Example 15**  To find the residue of the function

$$f(z) = \cot z$$

at its isolated singular point.

The function $f(z)$ has the isolated singular point $z = n\pi$ ($n=0, \pm 1, \pm 2, \cdots$), and each singular point is a simple pole, and $f(z)$ can be written as

$$f(z) = \cot z = \frac{\cos z}{\sin z}$$

where $p(z) = \cos z$ and $q(z) = \sin z$.
So

$$\operatorname*{Res}_{z=n\pi} f(z) = \frac{p(n\pi)}{q'(n\pi)} = \frac{(-1)^n}{(-1)^n} = 1$$

**Example 16**  To evaluate the integral

$$\int_C \tan \pi z \, dz$$

where $C$ is the circle $|z| = n$, $n$ is any positive integer, described in the counterclockwise directions.

The function $f(z)$ has the isolated singular point $z = \left(k + \frac{1}{2}\right)$ ($k=0, \pm 1, \pm 2, \cdots$), and each singular point is a simple pole, and $f(z)$ can be written as

$$f(z) = \tan \pi z = \frac{\sin \pi z}{\cos \pi z}$$

where $p(z) = \sin \pi z$ and $q(z) = \cos \pi z$
So

$$\operatorname*{Res}_{z=k+\frac{1}{2}} f(z) = \frac{p\left(k+\frac{1}{2}\right)}{q'\left(k+\frac{1}{2}\right)} = -\frac{1}{\pi}$$

Thus

$$\int_C \tan\pi z\, dz = 2\pi i \sum_{|k+\frac{1}{2}|<n} \underset{z=k+\frac{1}{2}}{\text{Res}}(\tan\pi z) = 2\pi i\left(-\frac{1}{\pi}n\right) = -4ni$$

## Exercises

1. Find the residues of the function

   (a) $\dfrac{z}{(z-1)(z+1)^2}$, $z=\pm 1$     (b) $\dfrac{1}{\sin z}$, $z=n\pi$, $(n=0, \pm 1, \cdots)$

   (c) $\dfrac{1-e^{2z}}{z^4}$, $z=0$     (d) $e^{\frac{1}{z-1}}$, $z=0$

   (e) $\dfrac{z^{2n}}{(z-1)^n}$, $z=1$     (f) $\dfrac{e^z}{z^2-1}$, $z=\pm 1$

2. Find the residue at $z=0$ of the function

   (a) $\dfrac{1}{z+z^2}$     (b) $z\cos\left(\dfrac{1}{z}\right)$     (c) $\dfrac{z-\sin z}{z}$

   (d) $\dfrac{\cos z}{z^4}$     (e) $\dfrac{\sh z}{z^4(1-z^2)}$     (f) $\dfrac{z^2+1}{z}$

   (g) $\dfrac{z^3}{(z-1)(z^4+2)}$     (h) $\dfrac{e^z}{z^2}$     (i) $\dfrac{\sin z}{z^7}$

   (j) $\dfrac{e^z}{\sin z}$     (k) $\dfrac{\log(1+z)}{z^2}$     (l) $\csc^2 z$

3. Find the residues of the following functions at 1.

   (a) $\dfrac{1}{(z^2-1)(z+2)}$     (b) $\dfrac{(z^3-1)(z+2)}{(z^4-1)^2}$

4. To evaluate the integral of each of these functions around the circle $|z|=3$ in the positive sense:

   (a) $\dfrac{e^{-z}}{z^2}$     (b) $\dfrac{e^{-z}}{(z-1)^2}$     (c) $z^2 e^{\frac{1}{z}}$     (d) $\dfrac{z+1}{z^2-2z}$

5. To evaluate the integral of each of these functions around the circle $|z|=2$ in the positive sense:

   (a) $\dfrac{z^5}{1-z^3}$     (b) $\dfrac{1}{1+z^2}$     (c) $\dfrac{1}{z}$

6. To evaluate the integral of each of these functions:

   (a) $\displaystyle\int_{|z|=1}\dfrac{dz}{z\sin z}$     (b) $\dfrac{1}{2\pi i}\displaystyle\int_{|z|=2}\dfrac{e^z}{1+z^2}dz$

   (c) $\displaystyle\int_C \dfrac{dz}{(z-1)^2(z^2+1)}$, $C: x^2+y^2=2(x+y)$

7. Show that the singular point of each of the following functions is a pole. Determine the order $m$ of that pole and the corresponding residue.

   (a) $\dfrac{1-\text{ch}z}{z^3}$     (b) $\dfrac{1-e^{2z}}{z^4}$     (c) $\dfrac{e^{2z}}{(z-1)^2}$

8. In each case, show that any singular point of the function is a pole. Determine the order $m$ of each pole, and find the corresponding residue.

   (a) $\dfrac{z^2+2}{z-1}$     (b) $\left(\dfrac{z}{2z+1}\right)^3$     (c) $\dfrac{e^z}{z^2+\pi^2}$

9. Find the value of the integral

$$\int_C \frac{3z^3+2}{(z-1)(z^2+9)}\,dz$$

   taken counterclockwise around the circle (a) $|z-2|=2$; (b) $|z|=4$.

10. Find the value of the integral

$$\int_C \frac{dz}{z^3(z+4)}$$

   taken counterclockwise around the circle (a) $|z|=2$; (b) $|z+2|=3$.

11. Evaluate the integral

$$\int_C \frac{\text{ch}\pi z}{z(z^2+1)}\,dz$$

   where $C$ is the circle $|z|=2$, described in the positive sense.

12. To evaluate the integral of $f(z)$ around the positively oriented circle $|z|=3$ when

   (a) $f(z)=\dfrac{(3z+2)^2}{z(z-1)(2z+5)}$     (b) $f(z)=\dfrac{z^3(1-3z)}{(1+z)(1+2z^4)}$

   (c) $f(z)=\dfrac{z^3 e^{\frac{1}{z}}}{1+z^3}$

13. Let $C$ denote the positively oriented circle $|z|=2$ and evaluate the integral

   (a) $\int_C \tan z\,dz$     (b) $\int_C \dfrac{dz}{\text{sh}2z}$

14. Find the integral

$$\int_C \frac{1}{z^4-1}\,dz$$

where $C$ is a circle of radius $\frac{1}{2}$ centered at $i$.

15. Find the integrals, where $C$ is the circle of radius 8 centered at the origin.

   (a) $\int_C \frac{1}{\sin z} dz$      (b) $\int_C \frac{1}{1-\cos z} dz$      (c) $\int_C \frac{1+z}{1-e^z} dz$

   (e) $\int_C \tan z \, dz$      (f) $\int_C \frac{1+z}{1-\sin z} dz$

16. Suppose that $f(z)$ has a simple pole at $z = z_0$ and let $g(z)$ be analytic in an open set containing $z_0$. Show that

$$\operatorname{Res}_{z=z_0}\{f(z)g(z)\} = g(z_0) \operatorname{Res}_{z=z_0} f(z)$$

17. Let $f(z)$ be analytic in the plane except for isolated singularities at $z_1$, $z_2$, $\cdots$, $z_m$. Show that

$$\operatorname{Res}_{z=\infty} f(z) = -\sum_{k=1}^{m} \operatorname{Res}_{z=z_k} f(z)$$

What can you say if $f(z)$ has infinitely many isolated singularities?

18. Let $z_1$, $z_2$, $\cdots$, $z_n$ be distinct complex numbers. Let $C$ be a circle around $z_1$ such that $C$ and its interior do not contain $z_j$ for $j > 1$. Let

$$f(z) = (z-z_1)(z-z_2)\cdots(z-z_n)$$

Find

$$\int_C \frac{1}{f(z)} dz$$

# 2 Applications of Residue

The calculus of residues provides a very efficient tool for the evaluation of certain types of definite and improper integrals occurring in real analysis.

## 2.1 The Type of Definite Integral $\int_0^{2\pi} F(\sin\theta, \cos\theta) d\theta$

**Theorem 5.2.1** Suppose certain definite integrals of the type

$$\int_0^{2\pi} F(\sin\theta, \cos\theta) d\theta \tag{5.2.1}$$

where $\theta$ varies from 0 to $2\pi$. Hence we write

$$z = e^{i\theta} \; (0 \leq \theta \leq 2\pi) \tag{5.2.2}$$

Formally,
$$\sin\theta = \frac{z - z^{-1}}{2i}, \quad \cos\theta = \frac{z + z^{-1}}{2}, \quad d\theta = \frac{dz}{iz} \qquad (5.2.3)$$
enable us to transform integral (5.2.1) into the contour integral
$$\int_{|z|=1} F\left(\frac{z - z^{-1}}{2i}, \frac{z + z^{-1}}{2}\right) \frac{dz}{iz} \qquad (5.2.4)$$
of a function of $z$ around the circle $|z|=1$ in the positive direction.

**Example 17** To evaluate the integral
$$\int_0^{2\pi} \frac{d\theta}{a + \sin\theta} (a > 1)$$

Let
$$z = e^{i\theta} (0 \leqslant \theta \leqslant 2\pi)$$

The integral takes the form
$$\int_{|z|=1} \frac{2}{z^2 + 2iaz - 1} dz$$

The only pole inside the circle $|z|=1$ is at
$$z_1 = -ia + i\sqrt{a^2 - 1}$$

Consequently,
$$\int_0^{2\pi} \frac{d\theta}{a + \sin\theta} (a > 1) = 2\pi i \operatorname*{Res}_{z=z_1} f(z) = 2\pi \frac{1}{\sqrt{a^2 - 1}}$$

**Example 18** Find the integral
$$\int_0^{2\pi} \frac{d\theta}{1 + a\sin\theta} (-1 < a < 1)$$

Let
$$z = e^{i\theta} (0 \leqslant \theta \leqslant 2\pi)$$

The integral takes the form
$$\int_{|z|=1} \frac{2/a}{z^2 + (2i/a)z - 1} dz$$

Because $|a| < 1$, the only pole inside the circle $|z|=1$ is at
$$z_1 = \left(\frac{-1 + \sqrt{1 - a^2}}{a}\right) i$$

If we write
$$f(z) = \frac{\phi(z)}{z - z_1} \text{ where } \phi(z) = \frac{2/a}{z - z_2}$$

Consequently,

$$\int_C \frac{2/a}{z^2 + (2i/a)z - 1} dz = 2\pi i \phi(z_1) = \frac{2\pi}{\sqrt{1-a^2}}$$

**Example 19** Find the integral

$$\int_0^{2\pi} \frac{d\theta}{1+\cos^2\theta}$$

Let

$$z = e^{i\theta} \ (0 \leqslant \theta \leqslant 2\pi)$$

The integral takes the form

$$\int_{|z|=1} \frac{4z}{i(z^4 + 6z^2 + 1)} dz$$

If we let $w = z^2$, the circle $|w| = 1$ is the same as $|z| = 1$, however, the circle $|w| = 1$ is traversed *twice* in the counterclockwise direction. So

$$\int_{|z|=1} \frac{4z}{i(z^4 + 6z^2 + 1)} dz = 2 \int_{|w|=1} \frac{2w}{i(w^2 + 6w + 1)} dw$$
$$= \frac{4}{i} \int_{|w|=1} \frac{w}{w^2 + 6w + 1} dw$$

Suppose

$$f(w) = \frac{w}{w^2 + 6w + 1}$$

Then the function $f(w)$ has the isolated singular point $w_1 = -3 + \sqrt{8}$ and $w_2 = -3 - \sqrt{8}$. The only pole inside the circle $|w| = 1$ is at $w_1 = -3 + \sqrt{8}$. Then

$$\operatorname*{Res}_{w=-3+\sqrt{8}} f(z) = \frac{1}{(w_1 - w_2)} = \frac{1}{4\sqrt{2}}$$

Such that

$$\int_0^{2\pi} \frac{d\theta}{1+\cos^2\theta} = 2\pi i \frac{4}{i} \frac{1}{4\sqrt{2}} = \sqrt{2}\pi$$

## 2.2 The Type of Improper Integral $\int_{-\infty}^{\infty} \frac{p(x)}{q(x)} dx$

**Theorem 5.2.2** An integral of the form $\int_{-\infty}^{\infty} \frac{p(x)}{q(x)} dx$ converges if and only if the rational function $\frac{p(z)}{q(z)}$, where the degree of the denominator is at least two units higher than the degree of the numerator, and if no pole lies on the real axis. Then

$$\int_{-\infty}^{\infty} \frac{p(x)}{q(x)} dx = 2\pi i \sum_{\substack{k=1 \\ \operatorname{Im} z_k > 0}}^{n} \operatorname*{Res}_{z=z_k} \frac{p(z)}{q(z)} \qquad (5.2.5)$$

**Example 20** To compute the integral

$$\int_{-\infty}^{+\infty} \frac{1}{x^4 + 1} dx$$

Let the function

$$f(z) = \frac{1}{x^4 + 1}$$

Its poles are at

$$c_k = e^{i\frac{\pi + 2k\pi}{4}}, \quad k = 0, 1, 2, 3$$

We find that all the poles of $f(z)$ are simple. There are two zeros in the upper half plane are $c_0$ and $c_1$. So we have

$$\int_{-\infty}^{+\infty} \frac{1}{x^4 + 1} dx = 2\pi i [\operatorname*{Res}_{z=c_0} f(z) + \operatorname*{Res}_{z=c_1} f(z)]$$

$$= 2\pi i \left( \frac{1}{4} e^{\frac{3}{4}\pi i} + \frac{1}{4} e^{\frac{1}{4}\pi i} \right) = \frac{\pi}{\sqrt{2}}$$

**Example 21** To evaluate the integral

$$\int_0^{\infty} \frac{x^2}{x^6 + 1} dx$$

Let the function

$$f(z) = \frac{z^2}{z^6 + 1}$$

Its poles are at

$$c_k = e^{i\frac{\pi(1+2k)}{6}} \quad (k = 0, 1, 2, \cdots, 5)$$

all the poles of $f(z)$ are simple. There are three zeros in the upper half plane are $c_0$, $c_1$ and $c_2$. We see that

$$\int_{-\infty}^{\infty} f(x) dx = 2\pi i (B_0 + B_1 + B_2)$$

where $B_k$ is the residue of $f(z)$ at $c_k$ ($k = 0, 1, 2$)

We find that

$$B_k = \operatorname*{Res}_{z=c_k} \frac{z^2}{z^6 + 1} = \frac{c_k^2}{6 c_k^5} = \frac{1}{6 c_k^3} \quad (k = 0, 1, 2)$$

Thus
$$\int_{-\infty}^{\infty} f(x)dx = \frac{\pi}{3}$$

Since the integrand here is even, we know that
$$\int_0^{\infty} \frac{x^2}{x^6+1}dx = \frac{\pi}{6}$$

**Example 22** To evaluate the integral
$$\int_{-\infty}^{\infty} \frac{x^4}{(2+3x^2)^4}dx$$

Let the function
$$f(z) = \frac{z^4}{(2+3z^2)^4}$$

Its poles are at $z_1 = \sqrt{\frac{2}{3}}i$ and $z_2 = -\sqrt{\frac{2}{3}}i$, all the poles of $f(z)$ are simple. Only the point $z = z_1$ is in the upper half plane. Let $t = z - z_1$, then $f(z)$ can be written as

$$f(z) = \frac{z^4}{(2+3z^2)^4} = \frac{z^4}{3^4(z-z_1)^4(z+z_1)^4} = \frac{(t+z_1)^4}{3^4 t^4 (t+2z_1)^4}$$

$$= \frac{1}{3^4 t^4} \cdot \frac{z_1^4 + 4z_1^3 t + 6z_1^2 t^2 + 4z_1 t^3 + t^4}{16 z_1^4 + 32 z_1^3 t + 24 z_1^2 t^2 + 8 z_1 t^3 + t^4}$$

$$= \frac{1}{3^4 t^4}\left(\frac{1}{16} + \frac{t}{8z_1} + \frac{t^2}{32 z_1^2} - \frac{t^3}{32 z_1^3} + \cdots \right)$$

We see that
$$\operatorname*{Res}_{z=z_1} f(z) = -\frac{1}{3^4 \cdot 32 z_1^3} = -\frac{i}{576\sqrt{6}}$$

So
$$\int_{-\infty}^{\infty} f(x)dx = 2\pi i \operatorname*{Res}_{z=\sqrt{\frac{2}{3}}i} f(z) = \frac{\pi}{288\sqrt{6}}$$

## 2.3 The Type of Improper Integral $\int_{-\infty}^{+\infty}\frac{p(x)}{q(x)}\sin ax\,dx$ or $\int_{-\infty}^{+\infty}\frac{p(x)}{q(x)}\cos ax\,dx$

**Theorem 5.2.3** Assume that $p(x)$ and $q(x)$ are polynomials with real coefficients and no factors in common. Also, $q(z)$ has no real zeros. Then

## Chapter V  Calculus of Residues

$$\int_{-\infty}^{\infty} \frac{p(x)}{q(x)} e^{iax} dx = 2\pi i \sum_{\substack{k=1 \\ \mathrm{Im} z_k > 0}}^{n} \mathrm{Res}_{z=z_k} \frac{p(z)}{q(z)} e^{iaz} \qquad (5.2.6)$$

**Example 23**  To find the value of the integral

$$\int_{-\infty}^{\infty} \frac{x \sin x}{x^2 + 2x + 2} dx$$

Let

$$f(z) = \frac{z}{z^2 + 2z + 2} = \frac{z}{(z - z_1)(z - \overline{z_1})}$$

its poles are at $z_1 = -1 + i$, $\overline{z_1} = -1 - i$. The point $z_1$ lies in the upper half plane, is a simple pole of the function $f(z)e^{iz}$, and

$$\mathrm{Res}_{z=z_1} f(z) = \frac{z_1 e^{iz_1}}{z_1 - \overline{z_1}}$$

So that

$$\int_{-\infty}^{\infty} \frac{x \sin x}{x^2 + 2x + 2} dx = \mathrm{Im}(2\pi i B_1) = \frac{\pi}{e} (\sin 1 + \cos 1)$$

**Example 24**  To find the value of the integral

$$\int_{0}^{\infty} \frac{\cos mx}{1 + x^2} dx$$

Let

$$f(z) = \frac{1}{1 + z^2} = \frac{1}{(z - z_1)(z - \overline{z_1})}$$

its poles are at $z_1 = i$, $\overline{z_1} = -i$. The point $z_1$ lies in the upper half plane, is a simple pole of the function $f(z)e^{imz}$, and

$$\mathrm{Res}_{z=z_1} f(z) e^{imz} = \frac{e^{imz_1}}{z_1 - \overline{z_1}}$$

So that

$$\int_{0}^{\infty} \frac{\cos mx}{1 + x^2} dx = \frac{1}{2} \int_{0}^{\infty} \frac{\cos mx}{1 + x^2} dx = \frac{1}{2} \mathrm{Re}[2\pi i \, \mathrm{Res}_{z=z_1} f(z) e^{imz}] = \pi i \frac{e^{-m}}{2i} = \frac{\pi e^{-m}}{2}$$

### Exercises

1. Find the value of the integral

(a) $\int_{0}^{2\pi} \frac{d\theta}{a + \cos\theta} (a > 1)$

(b) $\displaystyle\int_0^{2\pi} \frac{dx}{(2+\sqrt{3}\cos x)^2}$

(c) $\displaystyle\int_0^{2\pi} \frac{dx}{a+\sin^2 x}$, $|a|>1$

(d) $\displaystyle\int_0^{2\pi} \frac{1}{2-\sin x}dx$

(e) $\displaystyle\int_0^{2\pi} \frac{1}{(a+b\cos\theta)^2}d\theta$ for $0<b<a$

(f) $\displaystyle\int_0^{2\pi} \frac{\cos^2 3\theta d\theta}{5-4\cos 2\theta}$

(g) $\displaystyle\int_0^{2\pi} \frac{d\theta}{1+a\cos\theta}(-1<a<1)$

(h) $\displaystyle\int_0^{2\pi} \frac{1}{1+a^2-2a\cos\theta}d\theta$ if $0<a<1$

(i) $\displaystyle\int_0^{\pi} \frac{\cos 2\theta d\theta}{1-2a\cos\theta+a^2}(-1<a<1)$

(j) $\displaystyle\int_0^{\pi} \frac{d\theta}{(a+\cos\theta)^2}(a>1)$

(k) $\displaystyle\int_0^{\pi} \tan(\theta+ia)d\theta$ ($a$ is real number and $a\neq 0$)

(l) $\displaystyle\int_0^{\pi} \frac{1}{1+\sin^2\theta}d\theta$

(m) $\displaystyle\int_0^{\pi} \frac{1}{3+2\cos\theta}d\theta$

(n) $\displaystyle\int_0^{\frac{\pi}{2}} \frac{1}{(a+\sin^2\theta)^2}d\theta (a>0)$

(o) $\displaystyle\int_0^{\pi} \frac{a}{a^2+\sin^2\theta}d\theta = \int_0^{2\pi} \frac{a}{1+2a^2-\cos\theta}d\theta$

(p) $\displaystyle\int_0^{\pi} \sin^{2n}\theta d\theta (n=1,2,\cdots)$

2. Let $n$ be an even integer. Find
$$\int_0^{2\pi} (\cos x)^n dx$$
by the method of residues.

3. To evaluate the improper integrals

(a) $\displaystyle\int_0^{+\infty} \frac{x^2}{(x^2+1)(x^2+4)}dx$

(b) $\int_{-\infty}^{+\infty} \dfrac{x^2}{(x^2+a^2)^2}dx \, (a>0)$

(c) $\int_{0}^{\infty} \dfrac{dx}{(x^2+1)^2}$

(d) $\int_{0}^{\infty} \dfrac{x^2\,dx}{(x^2+9)(x^2+4)^2}$

(e) $\int_{-\infty}^{\infty} \dfrac{1}{1+x^6}dx$

(f) $\int_{-\infty}^{\infty} \dfrac{x^2}{x^4+1}dx$

(g) $\int_{-\infty}^{\infty} \dfrac{x-1}{x^5-1}dx$

(h) $\int_{-\infty}^{\infty} \dfrac{x\,dx}{(x^2+1)(x^2+2x+2)}$

4. Use residues to evaluate the integrals

(a) $\int_{-\infty}^{+\infty} \dfrac{\cos x\,dx}{(x^2+1)(x^2+9)}$

(d) $\int_{0}^{+\infty} \dfrac{x\sin mx}{x^4+a^4}dx, \, (m>0, a>0)$

(c) $\int_{-\infty}^{\infty} \dfrac{\cos ax}{x^2+1}dx \, (a>0)$

(d) $\int_{0}^{\infty} \dfrac{\cos ax}{(x^2+b^2)^2}dx \, (a>0, b>0)$

(e) $\int_{-\infty}^{\infty} \dfrac{x\sin x}{x^2+a^2}dx$

(f) $\int_{-\infty}^{\infty} \dfrac{\sin x\,dx}{x^2+4x+5}$

(g) $\int_{-\infty}^{\infty} \dfrac{(x+1)\cos x}{x^2+4x+5}dx$

5. To evaluate the improper integrals

(a) $\int_{0}^{\infty} \dfrac{\sin x}{x}dx$

(b) $\int_{0}^{\infty} \dfrac{\sin^2 x}{x^2}dx$

(c) $\int_{-\infty}^{\infty} \dfrac{\cos x}{a^2-x^2}dx$ for $a>0$

(d) $\int_{0}^{\infty} \dfrac{\sin x}{x(x^2+a^2)}dx$ for $a>0$

(e) $\int_0^\infty \dfrac{\sin x}{x(x^2+1)^2}dx$

(f) $\int_0^\infty \dfrac{\cos(ax)-\cos(bx)}{x^2}dx\,(a\geqslant 0, b\geqslant 0)$

6. Show that for a positive integer $n\geqslant 2$,

$$\int_0^\infty \frac{1}{1+x^n}dx = \frac{\dfrac{\pi}{n}}{\sin\dfrac{\pi}{n}}$$

7. Let $f(z)$ be an entire function and let $a$, $b\in C$ such that $|a|<R$ and $|b|<R$. If $v(t)=Re^{it}$, $0\leqslant t\leqslant 2\pi$, evaluate $\int_v \dfrac{f(z)}{(z-a)(z-b)}dz$. Use this result to give another proof of Liouville's Theorem.

# 3  Argument Principle

**Definition 5.3.1**  A function $f(z)$ is said to be meromorphic in a domain $D$ if it is analytic throughout $D$ except for possible poles.

**Theorem 5.3.1 (Argument Principle)**  Suppose that

(1) a function $f(z)$ is meromorphic in the domain interior to a positively oriented simple closed contour $C$;

(2) $f(z)$ is analytic and nonzero on $C$;

(3) counting multiplicities, $Z$ is the number of zeros and $P$ is the number of poles of $f(z)$ inside $C$.

Then

$$\frac{1}{2\pi i}\int_C \frac{f'(z)}{f(z)}dz = Z - P \qquad (5.3.1)$$

**Proof**  Suppose that the zeros of the function $f(z)$ inside $C$ are $a_k$, $(k=1, 2, \cdots, p)$ and the poles are $b_j$, $(j=1, 2, \cdots, q)$. We observe that the integrand $\dfrac{f'(z)}{f(z)}$ is analytic inside and on $C$ except at the points $a_k$, $(k=1, 2, \cdots, p)$ and $b_j$, $(j=1, 2, \cdots, q)$ inside $C$. Applying the residue theorem, then, we find that

$$\int_C \frac{f'(z)}{f(z)} dz = 2\pi i \left( \sum_{k=1}^{p} \operatorname*{Res}_{z=a_k} \frac{f'(z)}{f(z)} + \sum_{j=1}^{q} \operatorname*{Res}_{z=b_j} \frac{f'(z)}{f(z)} \right) = 2\pi i (Z-P) \quad (5.3.2)$$

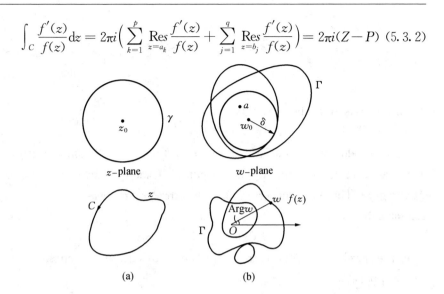

**Fig. 38**

**Theorem 5.3.2(Rouché)** Suppose that
(1) two functions $f(z)$ and $g(z)$ are analytic inside and on a simple closed contour $C$;
(2) $|f(z)| > |g(z)|$ at each point on $C$.
Then $f(z)$ and $f(z)+g(z)$ have the same number of zeros, counting multiplicities, inside $C$.

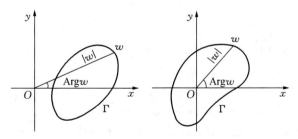

**Fig. 39**

**Example 25** To determine the number of roots of the equation
$$z^7 - 4z^3 + z - 1 = 0 \quad (5.3.3)$$
inside the circle $|z|=1$.
Hint: Look for the biggest term when $|z|=1$ and apply Rouché's theorem.

We write
$$f(z) = -4z^3$$
and
$$g(z) = z^7 + z - 1$$
Then observe that when $|z| = 1$, we have
$$|f(z)| = 4|z|^3 = 4 \text{ and } |g(z)| \leq |z|^7 + |z| + 1 = 3$$
The conditions in Rouché's theorem are thus satisfied. Consequently, since $f(z)$ has three zeros, counting multiplicities, inside the circle $|z| = 1$, so does $f(z) + g(z)$. That is, equation (5.3.3) has three roots there.

**Example 26** Let
$$P(z) = z^8 - 5z^3 + z - 2$$
Find the number of roots of this polynomial inside the unit circle.

Suppose
$$f(z) = -5z^3$$
and
$$g(z) = z^8 + z - 2$$
For $|z| = 1$ it is immediate that
$$|g(z)| = |z^8 + z - 2| \leq |f(z)| = 5$$
Hence $f(z)$ and $P(z)$ have the same number of zeros inside the unit circle, and this number is clearly equal to 3.

## Exercises

1. Let $C$ denote the unit circle $|z| = 1$, described in the positive sense. To determine the value of $\Delta_C \arg f(z)$ when
   (a) $f(z) = z^2$   (b) $f(z) = (z^3 + 2)/z$   (c) $f(z) = (2z - 1)^7/z^3$
2. Suppose that a function $f(z)$ is meromorphic in the domain $D$ interior to a simple closed contour $C$ on with $f(z)$ is analytic and nonzero, and let $D_0$ denote the domain consisting of all points in $D$ except for poles. Point out that if $f(z)$ is not identically equal to zero in $D_0$, then the zeros of $f(z)$ in $D$ are all of finite order and that they are finite in number.
3. Determine the number of zeros, counting multiplicities, of the polynomial
   (a) $z^6 - 5z^4 + z^3 - 2z$   (b) $2z^4 - 2z^3 + 2z^2 - 2z + 9$
   inside the circle $|z| = 1$.

4. Determine the number of zeros, counting multiplicities, of the polynomial
   (a) $z^4 + 3z^3 + 6$  (b) $z^4 - 2z^3 + 9z^2 + z - 1$  (c) $z^5 + 3z^3 + z^2 + 1$
   inside the circle $|z| = 2$.
5. Determine the number of zeros of the polynomial
   $$z^{87} + 36z^{57} + 71z^4 + z^3 - z + 1$$
   inside the circle
   (a) of radius 1;
   (b) of radius 2, centered at the origin.
6. Show that if $c$ is a complex number such that $|c| > e$, then the equation $cz^n = e^z$ has $n$ roots, counting multiplicities, inside the circle $|z| = 1$.
7. Let $a$ be real $>1$. Prove that the equation $ze^{a-z} = 1$ has a single solution with $|z| \leq 1$, which is real and positive.
8. Determine the number of zeros of the polynomial
   $$2z^5 - 6z^2 + z + 1$$
   in the annulus $1 \leq |z| \leq 2$.
9. How many roots of the equation $z^4 - 8z + 10 = 0$ have their modulus between 1 and 3?
10. How many roots of the equation
    $$z^4 + 8z^3 + 3z^2 + 8z + 3 = 0$$
    lie in the right half plane?
12. Let $f(z)$ be meromorphic on a domain $D$. Show that neither the poles nor the zeros of $f(z)$ have a limit point in $D$.
13. Let $\lambda > 1$ and show that the equation $\lambda - z - e^{-z} = 0$ has exactly one solution in the half plane $\{z: \text{Re} z > 0\}$. Show that this solution must be real. What happens to the solution as $\lambda \to 1$?
14. Let $f(z)$ be analytic in a neighborhood of a domain $D = \{z: |z| < 1\}$. If $|f(z)| < 1$ for $|z| = 1$, show that there is a unique $z$ with $|z| < 1$ and $f(z) = z$. If $|f(z)| \leq 1$ for $|z| = 1$, what can you say?
15. Let $f(z)$ be a function analytic inside and on the unit circle. Suppose that $|f(z) - z| < |z|$ on the unit circle.
    (a) Show that $\left| f'\left(\frac{1}{2}\right) \right| \leq 8$.
    (b) Show that $f(z)$ has precisely one zero inside of the unit circle.

# Chapter VI
## Conformal Mappings

In this chapter, we introduce and develop the concept of a conformal mapping. The purpose of this chapter is to study the special properties of mapping by analytic functions. Of all analytic functions the linear fractional transformations have the simplest mapping properties.

## 1  Analytic Transformation

### 1.1  Preservation of Domains of Analytic Transformation

A function $w=f(z)$ may be viewed as a mapping which represents a point $z$ by its image $w$.

**Theorem 6.1.1**  Suppose that $f(z)$ is analytic at $z_0$, $f(z_0)=w_0$, and that $f(z)-w_0$ has a zero of order $n$ at $z_0$. If $\varepsilon>0$ is sufficient small, there exists a corresponding $\delta>0$ such that for all $w$ with $|w-w_0|<\delta$ the equation $f(z)=w$ has exactly $n$ roots in the disk $|z-z_0|<\varepsilon$.

**Proof**  We can choose $\varepsilon$ so that $f(z)$ is defined and analytic for $|z-z_0|\leqslant\varepsilon$ and so that $z_0$ is the only zero of $f(z)-w_0$ in this disk. Let $\gamma$ be the circle $|z-z_0|=\varepsilon$ and $\Gamma$ its image under the mapping $w=f(z)$. Since $w_0$ belongs to the complement of the closed set $\Gamma$, there exists a neighborhood $|w-w_0|<\delta$ which does not intersect $\Gamma$. It follows immediately that all values $w$ in this neighborhood are taken the same number of times inside of $\gamma$. The equation $f(z)=w_0$ has exactly $n$ coinciding roots inside of $\gamma$, and hence every value $w$ is taken $n$ times. It is understood that multiple roots are counted according to their multiplicity, but if $\varepsilon$ is sufficiently small we can assert that all roots of the equation $f(z)=w$ are simple for $w\neq w_0$. Indeed, it is sufficient to choose $\varepsilon$ so that $f'(z)$ does not vanish for $0<|z-z_0|<\varepsilon$.

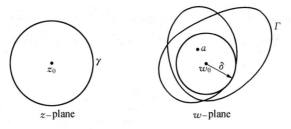

Fig. 40

**Corollary 6.1.2** A nonconstant analytic function maps open sets onto open sets.

This is merely another way of saying that imagine of every sufficiently small disk $|z-z_0|<\varepsilon$ contains a neighborhood $|w-w_0|<\delta$.

In the case $n=1$ there is one-to-one correspondence between the disk $|w-w_0|<\delta$ and an open subset $\Delta$ of $|z-z_0|<\varepsilon$.

Since open sets in the $z$-plane correspond to open sets in the $w$-plane the inverse function of $f(z)$ is continuous, and the mapping is topological. The mapping can be restricted to a neighborhood of $z_0$ contained in $\Delta$.

**Corollary 6.1.3** If $f(z)$ is analytic at $z_0$ with $f'(z_0)\neq 0$, it maps a neighborhood of $z_0$ conformally and topologically onto a region.

From the continuity of the inverse function it follows in the usual way that the inverse function is analytic, and hence the inverse mapping is likewise conformal. Conversely, if the local mapping is one to one, Theorem 6.1.1 can hold only with $n=1$, and hence $f'(z_0)$ must be different from zero.

## 1.2 Conformality of Analytic Transformation

Let $C$ be a smooth arc, represented by the equation
$$z=z(t) \quad (a\leqslant t\leqslant b)$$
and let $f(z)$ be a function defined at all points $z$ on $C$. The equation
$$w=f[z(t)] \quad (a\leqslant t\leqslant b)$$
is a parametric representation of the image $\Gamma$ of $C$ under the transformation $w=f(z)$.

Suppose that $C$ passes through a point $z_0=z(t_0)$ $(a\leqslant t_0\leqslant b)$ at which $f(z)$ is analytic and $f(z_0)\neq 0$. According to the chain rule theorem, then

$$w'(t_0) = f'[z(t_0)]z'(t_0) \qquad (6.1.1)$$

We interpret $z'(t)$ as a vector in the direction of a tangent vector at the point $z(t)$. This derivative $z'(t)$, if not zero, defines the direction of the curve at the point.

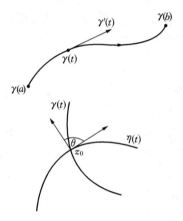

Fig. 41

**Definition 6.1.1** Consider $C_1: z=v(t)$ and $C_2: z=\eta(t)$ be two smooth arcs passing through the same point $z_0$, say
$$z_0 = v(t_0) = \eta(t_1)$$
Then the tangents vectors $v'(t_0)$ and $\eta'(t_1)$ determine an angle $\theta$ which is defined to be the angle between the curves.

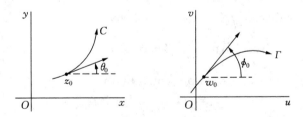

Fig. 42

Applying $f(z)$, the curves $\Gamma_1: w=f(v(t))$ and $\Gamma_2: w=f(\eta(t))$ pass through the point $f(z_0)$, and by the chain rule, tangent vectors of these image curves are
$$f'(z_0)v'(t_0) \quad \text{and} \quad f'(z_0)\eta'(t_1)$$

## Chapter VI  Conformal Mappings

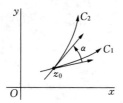
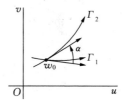

**Fig. 43**

**Theorem 6.1.4**  If $f(z)$ is analytic, $f'(z_0) \neq 0$ then the angle between the curves $C_1$, $C_2$ at the point $z_0$ is the same as the angle between the curves $\Gamma_1$, $\Gamma_2$ at the point $f(z_0)$.

**Proof**  Geometrically speaking, the tangent vectors under $f(z)$ are changed by multiplication with $f'(z_0)$, which can be represented in polar coordinates as a dilation and a rotation, so preserves the angles.

Because of this angle-preserving property, we introduce the following concept.

**Definition 6.1.2**  A map which preserves angles is called conformal mapping.

**Theorem 6.1.5**  A transformation $w = f(z)$ is conformal at a point $z_0$ if $f(z)$ is analytic there and $f'(z_0) \neq 0$.

**Definition 6.1.3**  A transformation $w = f(z)$, defined on a domain $D$, is referred to as a conformal transformation, or conformal mapping, when it is conformal at each point in $D$.

A related property of the mapping is derived by consideration of the modulus of $f'(z_0)$. We have

$$|f'(z_0)| = \left|\lim_{z \to z_0} \frac{f(z)-f(z_0)}{z-z_0}\right| = \lim_{z \to z_0} \frac{|f(z)-f(z_0)|}{|z-z_0|} \qquad (6.1.2)$$

Now $|z-z_0|$ is the length of a line segment joining $z_0$ and $z$, and $|f(z)-f(z_0)|$ is the length of the line segment joining the points $f(z_0)$ and $f(z)$ in the $w$ plane. Evidently, then, if $z$ is near the point $z_0$, the ratio

$$\frac{|f(z)-f(z_0)|}{|z-z_0|}$$

of the two lengths is approximately the number $f'(z_0)$. Note that the quantity $|f'(z_0)|$ represents an expansion if it is greater than unity and a contraction if it is less than unity.

**Theorem 6.1.6** If the mapping of $D$ by $w=f(z)$ is conformal transformation, then the inverse function $z=f^{-1}(w)$ is also analytic.

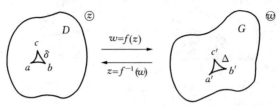

Fig. 44

**Proof** If $f'(z)\neq 0$, then the derivative of the inverse function must be equal to $\dfrac{1}{f'(z)}$ at the point $z=f^{-1}(w)$. If we write

$$z=x+iy,\quad z_0=x_0+iy_0 \quad\text{and}\quad f(z)=u(x,y)+iv(x,y)$$

We know that there is a neighborhood of the point $(x_0, y_0)$ throughout which the functions $u(x, y)$ and $v(x, y)$ along with their partial derivatives of all orders, are continuous.

We remark that the existence of the four partial derivatives is implied by the existence of $f'(z)$. We can write down the expressions for $f'(z)$

$$f'(z)=u_x+iv_x \tag{6.1.3}$$

For the quantity $|f'(z)|^2$ we have, for instance,

$$|f'(z)|^2 = u_x^2 + v_x^2 = u_x v_y - v_x u_y$$

Now the determinant

$$J=\begin{vmatrix} u_x & u_y \\ v_x & v_y \end{vmatrix} = u_x v_y - v_x u_y \tag{6.1.4}$$

which is known as the *Jacobian* of the transformation, is nonzero at the point $(x_0, y_0)$.

Since the transformation $w=f(z)$ is conformal at $z_0$. The above continuity conditions on the functions $u(x, y)$ and $v(x, y)$ and their derivatives, together with this condition on the Jacobian, are sufficient to ensure the existence of a local inverse at $(x_0, y_0)$. That is, if

$$u_0=u(x_0, y_0),\quad\text{and}\quad v_0=v(x_0, y_0) \tag{6.1.5}$$

Then from Mathematical Analysis, we know that there is a unique continuous transformation

$$x=x(u, v),\quad y=y(u, v) \tag{6.1.6}$$

defined on a neighborhood $N$ of the point $(u_0, v_0)$ and mapping that onto $(x_0, y_0)$.

### Exercises

1. Find the image of $\{z: \operatorname{Re} z < 0, |\operatorname{Im} z| < \pi\}$ under the exponential function.
2. Find the image of $\{z: |\operatorname{Im} z| < \frac{\pi}{2}\}$ under the exponential function.
3. Discuss the mapping properties of $\cos z$ and $\sin z$.
4. Discuss the mapping properties of $z^n$ and $z^{\frac{1}{n}}$ for $n > 2$.
5. Let $D$ be a region and suppose that $f: D \to C$ is analytic such that $f(D)$ is a subset of a circle. Show that $f(z)$ is constant.

## 2 Rational Functions

### 2.1 Polynomials

**Definition 6.2.1** Suppose that the polynomial
$$P(z) = c_0 + c_1 z + \cdots + c_n z^n \quad (c_n \neq 0) \tag{6.2.1}$$
The polynomial is then said to be of degree $n$.

In light of the preceding proposition, entire functions can be considered as polynomials of "infinite degree".

Because every constant is an analytic function with the derivatives zero, and the sum and product of two analytic functions are again analytic, it follows that every polynomial is an analytic function. Its derivatives is
$$P'(z) = c_1 + 2c_2 z + \cdots + nc_n z^{n-1} \tag{6.2.2}$$

Returning to the discussion of polynomials, we have that the multiplicity of a zero of a polynomial must be less than the degree of the polynomial.

For $n > 0$ the equation $P(z) = 0$ has at least one root. This is the so-called fundamental theorem of algebra which we have proved in Chapter 3.

**Theorem 6.2.1** If $P(z)$ is a polynomial and $z_1, z_2, \cdots, z_m$ are its zeros with $z_i$ having multiplicity $k_i$, then
$$P(z) = c_n (z - z_1)^{k_1} (z - z_2)^{k_2} \cdots (z - z_m)^{k_m}$$
and $k_1 + k_2 + \cdots + k_m$ is the degree of $P(z)$.

We also have a complete factorization
$$P(z) = c_n(z-z_1)(z-z_2)\cdots(z-z_n) \tag{6.2.3}$$

**Lemma 6.2.2** A directed line $z = a + bt$ determines a right half plane consisting of all points $z$ with
$$\operatorname{Im} \frac{z-a}{b} < 0$$

**Theorem 6.2.3 (Lucas's Theorem)** If all zeros of a polynomial $P(z)$ lies in a half plane, then all zeros of the derivative $P'(z)$ lie in the same half plane.

**Proof** From (6.2.3), we obtain
$$\frac{P'(z)}{P(z)} = \frac{1}{z-z_1} + \cdots + \frac{1}{z-z_n} \tag{6.2.4}$$

Suppose that the half plane $H$ is defined as the part of the plane where
$$\operatorname{Im} \frac{z-z_1}{b} < 0$$

If $z_k$ is in $H$ and $z$ is not, we have then
$$\operatorname{Im} \frac{z-z_k}{b} = \operatorname{Im} \frac{z-a}{b} - \operatorname{Im} \frac{z_k-a}{b} > 0$$

But the imaginary parts of reciprocal numbers have opposite sign. Therefore, under the same assumption,
$$\operatorname{Im} \frac{b}{z-z_k} < 0$$

If this is true for all $k$ that
$$\operatorname{Im} \frac{bP'(z)}{P(z)} = \sum_{k=1}^{n} \operatorname{Im} \frac{b}{z-z_k} < 0$$

In a shaper formulation the theorem tells us that the smallest convex polygon that contains the zeros of $P(z)$ also contains the zeros of $P'(z)$.

## 2.2 Rational Functions

**Definition 6.2.2** Suppose that the two polynomial $P(z)$ and $Q(z)$ have no common factors and hence no common zeros. We call the notation
$$f(z) = \frac{P(z)}{Q(z)}$$
be a rational function.

## Theorem 6.2.4

(1) $f(z)$ will be given the value $\infty$ at the zeros of $Q(z)$;

(2) $f'(z)$ has the same poles as $f(z)$ and the number of poles is the same;

(3) If $\alpha$ is any constant, the function $f(z)-\alpha$ has the same poles as $f(z)$, and consequently the same order;

(4) A rational function $f(z)$ of order $n$ has $n$ zeros and $n$ poles, and every equation $f(z)=\alpha$ has exactly $n$ roots.

**Definition 6.2.3** A rational function of order 1 is a linear fraction

$$S(z) = \frac{az+b}{cz+d} \quad (ad-bc \neq 0) \tag{6.2.5}$$

Such linear fraction also called linear fractional transformations. If $a$, $b$, $c$ and $d$ also satisfy $ad-bc \neq 0$ then $S(z)$ is called a Mobius transformation.

**Definition 6.2.4** If $S(z)$ is a Mobius transformation then

$$z = S^{-1}(w) = \frac{dw-b}{-cw+a} \tag{6.2.6}$$

The transformation $S^{-1}(z)$ is the inverse mapping of $S(z)$. The transformations $S(z)$ and $S^{-1}(z)$ are inverse to each other.

**Theorem 6.2.5** If $f(z)$ and $g(z)$ are both linear fractional transformations then it follows that $f[g(z)]$ is also.

**Definition 6.2.5** If $S(z)=z+\alpha$ then $S(z)$ is called a translation; if $S(z)=az$ with $a \neq 0$, then $S(z)$ is a dilation; if $S(z)=e^{i\theta}z$ then it is a rotation; finally, if $S(z)=\dfrac{1}{z}$ it is the inversion.

**Theorem 6.2.6** If $S(z)$ is a Mobius transformation then $S(z)$ is the composition of translations, dilations, and the inversion (Of course, some of these may be missing.)

**Definition 6.2.6** A fixed point of a transformation $w=f(z)$ is a point $z_0$ such that $f(z_0)=z_0$.

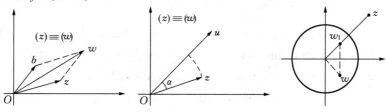

Fig. 45

**Theorem 6.2.7** Every fractional linear transformation, with the exception of the identity transformation $w=z$, has at most two fixed points in the extended plane.

### Exercises

1. If $S(z)=\dfrac{az+b}{cz+d}$ show that $S(R)=R$ if we can choose $a$, $b$, $c$, $d$ to be real numbers.

2. If $S(z)=\dfrac{az+b}{cz+d}$ find necessary and sufficient conditions that $S(C)=C$ where $C$ is the unit circle $\{z:|z|=1\}$.

3. Find the fixed points of the linear transformations

   (a) $w=\dfrac{z}{2z-1}$    (b) $w=\dfrac{2z}{3z-1}$    (c) $w=\dfrac{3z-4}{z-1}$

   (d) $w=\dfrac{z}{2-z}$    (d) $w=\dfrac{z-1}{z+1}$    (e) $w=\dfrac{6z-9}{z}$

4. Find the fixed points of a dilation, a translation and the inversion.

5. Let $S(z)=\dfrac{az+b}{cz+d}$ and $T(z)=\dfrac{\alpha z+\beta}{\gamma z+\delta}$; show that $S=T$ if there is a non zero complex number $\lambda$ such that $\alpha=\lambda a$, $\beta=\lambda b$, $\gamma=\lambda c$, $\delta=\lambda d$.

6. Let $T$ be a Mobius transformation with fixed points $z_1$ and $z_2$. If $S$ is a Mobius transformation show $S^{-1}TS$ has fixed points $S^{-1}z_1$ and $S^{-1}z_2$.

7. (a) Show that a Mobius transformation has $0$ and $\infty$ as its only fixed points if it is a dilation.
   (b) Show that a Mobius transformation has $\infty$ as its only fixed point if it is a translation.

8. Show that a Mobius transformation $T$ satisfies $T(0)=\infty$ and $T(\infty)=0$ if $T(z)=\dfrac{a}{z}$ for some $a$ in $C$.

9. Let $T$ be a Mobius transformation, $T\ne$ the identity. Show that a Mobius transformation $S$ commutes with $T$ if $S$ and $T$ have the same fixed points.

10. Show that a composition of two linear fractional transformations is again a linear fractional transformation.

11. Prove that if the origin is a fixed point of a fractional linear transforma-

Chapter VI  Conformal Mappings

tion, then the transformation can be written in the form $w=z/(ca+d)$, where $d\neq 0$.

12. Show that there is only one fractional linear transformation that maps three given distinct points $z_1$, $z_2$, and $z_3$ in the extended $z$ plane onto three specified distinct points $w_1$, $w_2$, and $w_3$ in the extended $w$ plane.

13. Prove that if a fractional linear transformation maps the points of the $x$ axis onto points of the $u$ axis, then the coefficients in the transformation are all real, except possibly for a common complex factor. The converse statement is evident.

14. Let $T(z)=(az+b)/(cz+d)$, where $ad-bc\neq 0$, be any fractional linear transformation other than $T(z)=z$. Show that $T^{-1}=T$ if and only if $d=-a$.

15. Prove that the reflection is not a linear transformation.

16. If
$$T_1(z)=\frac{z+2}{z+3}, \qquad T_2(z)=\frac{z}{z+1}$$
find $T_1T_2(z)$, $T_2T_1z$ and $T_1^{-1}T_2z$.

## 3  Fractional Linear Transformations

Consider the fractional linear transformation (6.2.5), where $a$, $b$, $c$, and $d$ are complex constants.

We define

(1) $S(\infty)=\dfrac{a}{c}$  if $c\neq 0$;

(2) $S(\infty)=\infty$  if $c=0$;

(3) $S\left(-\dfrac{d}{c}\right)=\infty$  if $c\neq 0$.

**Theorem 6.3.1**  Let $S(z)$ be a fractional linear transformation. If $S(z)$ has three fixed points, then $S(z)$ is the identity.

**Theorem 6.3.2**  A fractional linear transformation maps straight lines and circles onto straight lines and circles.

There is always a fractional linear transformation that maps three given distinct points $z_1$, $z_2$ and $z_3$ onto three specified distinct points $w_1$, $w_2$ and

$w_3$, respectively. Then the image $w$ of a point $z$ under such a transformation is given implicitly in terms of $z$. We illustrate here a more direct approach to finding the desired transformation.

**Definition 6.3.1** The cross ratio $(z_1, z_2, z_3, z_4)$ is the image of $z_1$ under the linear transformation which takes $z_2$ to 1, $z_3$ to 0 and $z_4$ to $\infty$.

**Theorem 6.3.3** If $z_1, z_2, z_3, z_4$ are distinct points in the extended plane and $w=S(z)$ is any linear transformation; then $w=S(z)$ is a Mobius map. If

$$(w_1, w_2, w_3, w_4) = (z_1, z_2, z_3, z_4) \qquad (6.3.1)$$

**Proof** Let $S(z_1) = (z_1, z_2, z_3, z_4)$; then $w=S(z)$ is a Mobius map. If $M = ST^{-1}$ then $M(Tz_2)=1$, $M(Tz_3)=0$, $M(Tz_4)=\infty$; hence, $ST^{-1}z_1 = (Tz_1, Tz_2, Tz_3, Tz_4)$. The desired result follows.

In the case that all specified points are all finite, we have

$$\frac{(w-w_1)(w_2-w_3)}{(w-w_3)(w_2-w_1)} = \frac{(z-z_1)(z_2-z_3)}{(z-z_3)(z_2-z_1)} \qquad (6.3.2)$$

**Theorem 6.3.4** If $z_2, z_3, z_4$ are distinct points in the extended complex plane and $w_2, w_3, w_4$ are also distinct points of extended complex plane, then there is one and only one Mobius transformation $w=S(z)$ such that $w_2=S(z_2)$, $w_3=S(z_3)$, $w_4=S(z_4)$.

**Proof** Let $T(z)=(z, z_2, z_3, z_4)$, $M(z)=(z, w_2, w_3, w_4)$ and put $S=M^{-1}T$. Clearly $S$ has the desired property. If $R$ is another Mobius map with $Rz_j = w_j$ for $j=2, 3, 4$ then $R^{-1}S$ has three fixed points $z_2, z_3, z_4$. Hence $R^{-1}S=I$, or $S=R$.

Equation (6.3.2) can be used to find the linear transformation in special cases.

**Example 1** Find the special case of transformation (6.2.5) that maps the points

$$z_1=1, \quad z_2=i \text{ and } z_3=-1$$

onto the points

$$w_1=i, \quad w_2=-1 \text{ and } w_3=1$$

By the formula,

$$\frac{(w-i)(-1-1)}{(w-1)(-1-i)} = \frac{(z-1)(i+1)}{(z+1)(i-1)}$$

We can solve for $w$ in terms of $z$ to give

$$w = \frac{z(1+2i)+1}{z+(1-2i)}$$

**Example 2** Suppose that the points
$$z_1 = 1, \ z_2 = 0 \text{ and } z_3 = -1$$
are mapped onto
$$w_1 = i, \ w_2 = \infty \text{ and } w_3 = 1$$
under (6.2.5). Find the coefficients $a$, $b$, $c$ and $d$ of (6.2.5).

Since $w_2 = \infty$ corresponds to $z_2 = 0$, we require that $d = 0$ in expression (6.2.5); and so
$$w = \frac{az+b}{cz} \quad (bc \neq 0)$$

Because 1 is to be mapped onto $i$ and $-1$ onto 1, we have the relations
$$ic = a+b \text{ and } -c = -a+b$$
and it follows that
$$b = \frac{i-1}{2}c, \quad a = \frac{i+1}{2}c$$

Finally, if we take $c = 2$, then we obtain the desired fractional linear transformation:
$$w = \frac{(i+1)z+(i-1)}{2z}$$

It is well known that three points in the plane determine a circle. A straight line in the plane will be called a circle. The next result explains when four points lie on a circle.

**Theorem 6.3.5** The cross ratio $(z_1, z_2, z_3, z_4)$ is real if and only if the four points lie on a circle or a straight line.

**Proof** Let $S$ be defined by $S(z) = (z, z_2, z_3, z_4)$. Let $S(z) = \frac{az+b}{cz+d}$; if $z = x \in \mathbf{R}$ and $w = S^{-1}(x)$ then $x = Sw$ implies that $S(w) = \overline{S(w)}$. That is,
$$\frac{aw+b}{cw+d} = \frac{\overline{a}\overline{w}+\overline{b}}{\overline{c}\overline{w}+\overline{d}}$$

Cross multiplying this gives
$$(a\overline{c}-\overline{a}c)|w|^2+(a\overline{d}-\overline{b}c)w+(\overline{b}c-\overline{d}a)\overline{w}+(b\overline{d}-\overline{b}d)=0 \quad (6.3.3)$$
If $a\overline{c}$ is real then $a\overline{c}-\overline{a}c = 0$; putting $\alpha = 2(a\overline{d}-\overline{b}c)$, $\beta = i(b\overline{d}-\overline{b}d)$ and multiplying (6.3.3) by $i$ gives

$$0 = \text{Im}(\alpha w) - \beta = \text{Im}(\alpha w - \beta) \qquad (6.3.4)$$

since $\beta$ is real. That is, $w$ lies on the line determined by (6.3.4) for fixed $\alpha$ and $\beta$. If $a\bar{c}$ is not real then (6.3.3) becomes

$$|w|^2 + \bar{\gamma}w + \gamma\bar{w} - \delta = 0$$

for some constants $\gamma$ in $C$, $\delta$ in $R$. Hence,

$$|w + \gamma| = \lambda$$

where

$$\lambda = \sqrt{|\gamma|^2 + \delta} = \left|\frac{ad - bc}{ac - \overline{ac}}\right| > 0$$

Since $\gamma$ and $\lambda$ are independent of $x$ and since (6.3.4) is the equation of a circle, the proof is finished.

**Theorem 6.3.6** A Mobius transformation takes circles onto circles.

**Proof** Let $\gamma$ be any circle in the extended complex plane and let $S$ be any Mobius transformation. Let $z_2$, $z_3$, $z_4$ be three distinct points on $\gamma$ and put $w_j = Sz_j$ for $j = 2, 3, 4$. Then $w_2$, $w_3$, $w_4$ determine a circle $\Gamma$. We claim that $S(\gamma) = \Gamma$. In fact, for any $z$ in the extended complex plane

$$(z, z_2, z_3, z_4) = (Sz, w_2, w_3, w_4) \qquad (6.3.5)$$

by Theorem 6.3.3. By the preceding proposition, if $z$ is on $\Gamma$ then both sides of (6.3.5) are real. But this says that $S(z) \in \Gamma$.

**Theorem 6.3.7** For any given circles $\gamma$ and $\Gamma$ in the extended complex plane there is a Mobius transformation $S$ such that $S(\gamma) = \Gamma$. Furthermore we can specify that $S$ take any three points on $\gamma$ onto any three points of $\Gamma$. If we do specify $Sz_j$ for $j = 2, 3, 4$ (distinct $z_j$ in $\gamma$) then $S$ is unique.

**Proof** Now let $\gamma$ and $\Gamma$ be two circles in the extended complex plane, and let $z_2, z_3, z_4 \in \gamma$; $w_2, w_3, w_4 \in \Gamma$. Put $R(z) = (z, z_2, z_3, z_4)$, $S(z) = (z, z_2, z_3, z_4)$. Then $T = S^{-1}R$ maps $\gamma$ onto $\Gamma$. In fact, $Tz_j = w_j$ for $j = 2, 3, 4$ and, as in the above proof, it follows that $T(\gamma) = \Gamma$.

The uniqueness part is a trivial exercise for the reader.

Now that we know that a Mobius map takes circles to circles, the next question is: What happens to the inside and the outside of these circles? To answer this we introduce some new concepts.

**Definition 6.3.2** The points $z$ and $z^*$ are said to be symmetric with respect

Chapter VI  Conformal Mappings

to the circle $C$ through $z_2$, $z_3$, $z_4$ if and only if
$$(z^*, z_2, z_3, z_4) = \overline{(z, z_2, z_3, z_4)} \qquad (6.3.6)$$

The two points $z$ and $\bar{z}$ are symmetric with respect to the real axis.

The point on $C$ and only those, are symmetric to themselves. The mapping which carries $z$ into $z^*$ is a one-to-one correspondence and is called reflection with respect to $C$.

Consider now the case of a finite circle $C$ of center $a$ and radius $R$, then $z$ and $z^*$ satisfy the relation
$$(z^* - a)(\bar{z} - \bar{a}) = R^2 \qquad (6.3.7)$$

**Theorem 6.3.8 (The symmetry Principle)**  If a Mobius transformation $S$ carries a circle $\gamma$ into a circle $\Gamma$, then it transforms any pair of symmetric points with respect to $\gamma$ into a pair of symmetric points with respect to $\Gamma$.

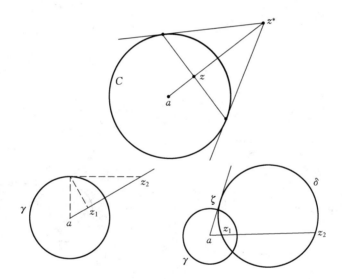

Fig. 46

**Proof**  Let $z_2$, $z_3$, $z_4 \in \gamma$; it follows that if $z$ and $z^*$ are symmetric with respect to $\gamma$ then
$$(Sz^*, Sz_2, Sz_3, Sz_4) = (z^*, z_2, z_3, z_4)$$
$$= \overline{(z, z_2, z_3, z_4)}$$
$$= \overline{(Sz, Sz_2, Sz_3, Sz_4)}$$

by Theorem 6.3.3. Hence $S(z)$ and $S(z^*)$ are symmetric with respect to $\Gamma$.

Now we will discuss orientation for circles in the extended complex plane; this will enable us to distinguish between the "inside" and "outside" of a circle.

**Definition 6.3.3**  If $\gamma$ is a circle then an orientation for $\gamma$ is an ordered triple of points $(z_1, z_2, z_3)$ such that each $z_j$ is in $\gamma$.

Intuitively, these three points give a direction to $\gamma$. That is we "go" from $z_1$ to $z_2$ to $z_3$. If only two points were given, this would, of course, be ambiguous.

**Theorem 6.3.9 (Orientation Principle)**  Let $\gamma$ and $\Gamma$ be two circles in the extended complex plane and let $S$ be a Mobius transformation such that $S(\gamma) = \Gamma$. Let $(z_1, z_2, z_3)$ be an orientation for $\gamma$. Then $S$ take the right side and the left side of $\gamma$ onto the right side and left side of $\Gamma$ with respect to the orientation $(Sz_1, Sz_2, Sz_3)$.

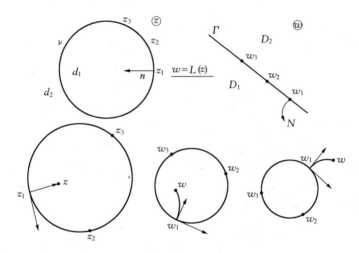

**Fig. 47**

### Exercises

1. Show that the definition of symmetry (6.3.2) does not depend on the choice of points $z_2, z_3, z_4$. That is, show that if $w_2, w_3, w_4$ are also in $C$ then equation (6.3.3) is satisfied if $(z^*, w_2, w_3, w_4) = (z, w_2, w_3, w_4)$.

2. Find the fractional linear transformation that maps the points $z_1=2$, $z_2=i$, $z_3=-2$ onto the points $w_1=1$, $w_2=i$, $w_3=-1$.
3. Find the fractional linear transformation that maps the points $z_1=-i$, $z_2=0$, $z_3=i$ onto the points $w_1=-1$, $w_2=i$, $w_3=1$. Into what curve is the imaginary axis $x=0$ transformed?
4. Find the fractional linear transformation that maps the points $z_1=\infty$, $z_2=i$, $z_3=0$ onto the points $w_1=0$, $w_2=i$, $w_3=\infty$.
5. Find the fractional linear transformation that maps distinct points $z_1$, $z_2$, $z_3$ onto the points $w_1=0$, $w_2=1$, $w_3=\infty$.
6. Lf $z_1$, $z_2$, $z_3$, $z_4$ are points on a circle, show that $z_1$, $z_3$, $z_4$ and $z_2$, $z_3$, $z_4$ determine the same orientation if and only if $(z_1, z_2, z_3, z_4)>0$.
7. Lf $S(z)=\dfrac{az+b}{cz+d}$ find $z_2$, $z_3$, $z_4$ (in terms of $a$, $b$, $c$, $d$) such that $Tz=(z, z_2, z_3, z_4)$.

## 4 Elementary Conformal Mappings

One of the most important problems is to determine the conformal mappings of one region onto another. In this section we shall consider those mappings which can be defined by elementary functions.

**Theorem 6. 4. 1** Transformation

$$w=\frac{az+b}{cz+d} \tag{6.4.1}$$

where $a$, $b$, $c$, $d$ is real numbers and $ad-bc>0$, maps the upper half plane Im $z>0$ onto the upper half plane Im $w>0$ and the boundary Im $z=0$ onto the boundary Im $w=0$.

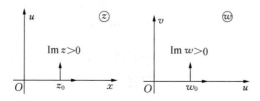

**Fig. 48**

**Example 3** The transformation

$$w=\frac{z-1}{z+1}$$

maps the half plane Im $z>0$ onto the half plane Im $w>0$ and the boundary Im $z=0$ onto the boundary Im $w=0$.

**Theorem 6.4.2** Transformation

$$w=e^{i\alpha}\frac{z-z_0}{z-\overline{z_0}} \quad (\text{Im } z_0>0) \tag{6.4.2}$$

where $\alpha$ is real, maps the upper half plane Im $z>0$ onto the open disk $|w|<1$ and the boundary Im $z=0$ onto the boundary $|w|=1$.

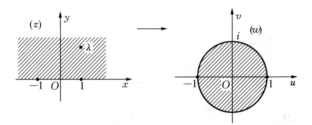

**Fig. 49**

**Example 4** Transformation

$$w=\frac{z-i}{z+i}$$

maps the upper half plane Im $z>0$ onto the open disk $|w|<1$ and the boundary Im $z=0$ onto the boundary $|w|=1$.

**Example 5** Transformation

$$w=-\frac{z-i}{z+i}$$

maps the upper half plane Im $z>0$ onto the open disk $|w|<1$, the boundary Im $z=0$ onto the boundary $|w|=1$, and $w(i)=0$, $\arg w'(i)=\frac{\pi}{2}$.

**Theorem 6.4.3** Let $f(z_0)=0$. Transformation

$$w=e^{i\alpha}\frac{z-z_0}{1-\overline{z_0}z}, \quad |z_0|<1 \tag{6.4.3}$$

where $\alpha$ is real, maps the open disc $|z|<1$ onto the open disc $|w|<1$ and the boundary $|z|=1$ onto the boundary $|w|=1$.

## Chapter VI  Conformal Mappings

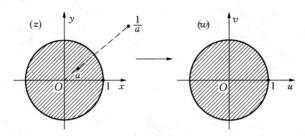

Fig. 50

**Corollary 6.4.4** If $f(0)=0$, which leaves the origin fixed, then $w=e^{i\alpha}z$ for some real number $\alpha$, so $w$ is a rotation.

**Example 6**  Transformation

$$w=2\frac{z-1/2}{z-2}$$

maps the open disc $|z|<1$ onto the open disc $|w|<1$, the boundary $|z|=1$ onto the boundary $|w|=1$, and satisfy the condition $\omega\left(\frac{1}{2}\right)=0$, $\omega(1)=-1$.

**Example 7**

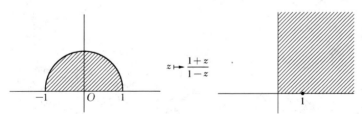

Upper half disc with first quadrant

Fig. 51

**Example 8**  Transformation

$$w=2(1+i)\frac{z-2i}{z+2i}$$

maps the upper half plane Im $z>0$ onto the open disk $|w-2i|<2$, the boundary Im $z=0$ onto the boundary $|w-2i|=2$, and $\omega(2i)=2i$, $\arg \omega'(2i)=-\frac{\pi}{2}$.

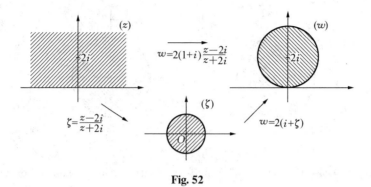

Fig. 52

**Example 9** Mappings by $z^n$ and Branches of $\sqrt[n]{z}$.

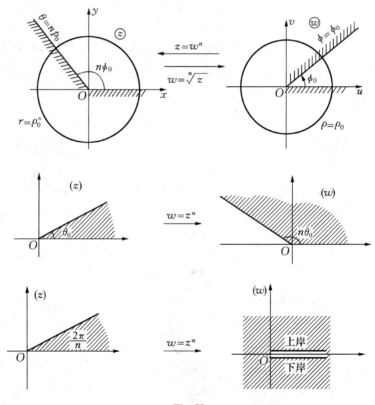

Fig. 53

## Example 10

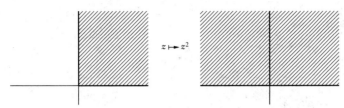

*Isomorphism between first quadrant and upper half plane*

**Fig. 54**

## Example 11

**Fig. 55**

## Example 12

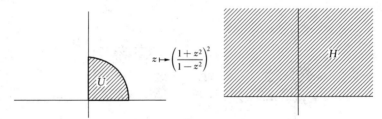

*Quarter disc with upper half plane*

**Fig. 56**

**Example 13** Mappings by $\operatorname{Ln} z$ and Branches of $e^z$.

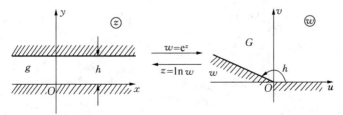

**Fig. 57**

**Example 14**

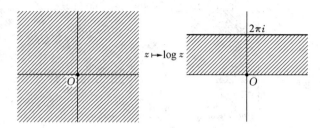

Fig. 58

**Example 15**

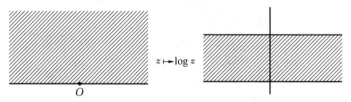

Upper half plane with a full strip

Fig. 59

**Example 16**

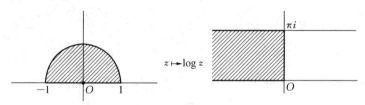

Upper half disc with a half strip

Fig. 60

**Example 17**

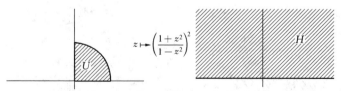

Quarter disc with upper half plane

Fig. 61

## Example 18

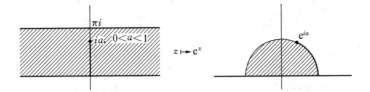

**Fig. 62**

## Example 19

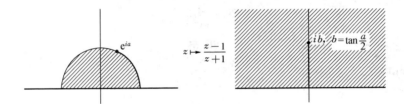

**Fig. 63**

## Example 20

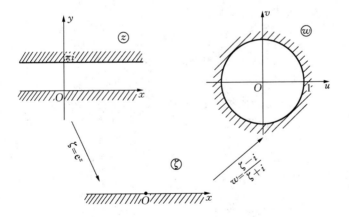

**Fig. 64**

**Example 21**

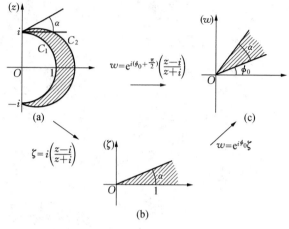

**Example 22**

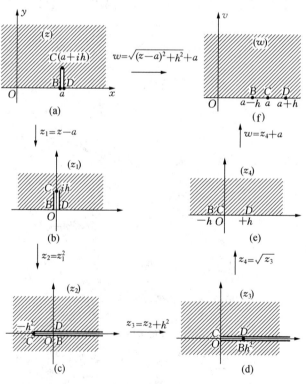

Fig. 66

## Example 23

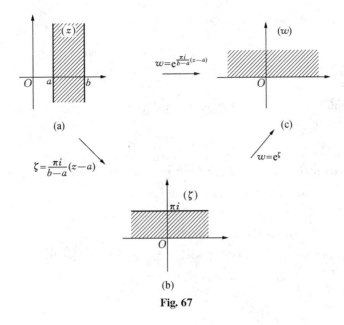

Fig. 67

## Exercises

1. Show that any linear transformation which transforms the real axis into itself can be written with real coefficients.
2. State why the transformation $w=iz$ is a rotation of the $z$ plane through the angle $\frac{\pi}{2}$.
   Then find the image of the infinite strip $0<x<1$.
3. Show that the transformation $w=iz+i$ maps the half plane $x>0$ onto the half plane $v>1$.
4. Find the region onto which the half plane $y>0$ is mapped by the transformation $w=(1+i)z$ by using
   (a) polar coordinates ; (b) rectangular coordinates. Sketch the region.
5. Find the image of the half plane $y>1$ under the transformation $w=(1-i)z$. Find the image of the semi-infinite strip $x>0$, $0<y<2$ when $w=iz+i$. Sketch the strip and its image.
6. (a) By finding the inverse of the transformation

$$w = \frac{i-z}{i+z}$$

and show that the transformation

$$w = i\frac{1-z}{1+z}$$

maps the disk $|z| \leqslant 1$ onto the half plane $w \geqslant 0$.

(b) Show that the linear fractional transformation $w = \frac{z-2}{z}$ can be written

$$Z = z-1, \quad W = i\frac{1-Z}{1+Z}, \quad w = iW$$

Then, with the aid of the result in part(a), verify that it maps the disk $|z-1| \leqslant 1$ onto the left half plane $\mathrm{Re}\, w \leqslant 0$.

7. Show that any linear transformation which transforms the real axis into itself can be written with real coefficients.

8. What is the image of the half strips as shown on the figure, under the mapping $z \to iz$? Under the mapping $z \to -iz$?

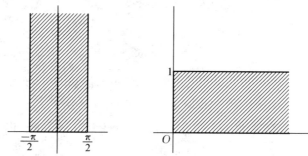

9. Let $\alpha$ be a real number, $0 \leqslant \alpha < 1$. Let $U_\alpha$ be the open set obtained from the unit disc by deleting the segment $[\alpha, 1]$, as shown on the figure.
   (a) Find an isomorphism of $U_\alpha$ with the unit disc from which the segment $[0, 1]$ has been deleted.

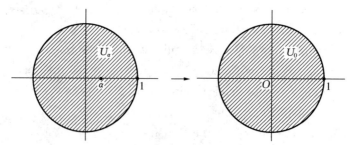

(b) For an isomorphism of $U_a$ with the upper half of the disc. Also find an isomorphism of $U_a$ with this upper half disc?

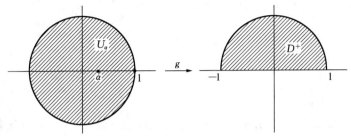

10. Map $D = C - \{z : -1 < z < 1\}$ onto the open unit disk for an analytic function $f(z)$. Can $f(z)$ be one-one?

## 5 The Riemann Mapping Theorem

**Theorem 6.5.1 (Lemma of Schwarz)** If $f(z)$ is analytic for $|z| < 1$ and satisfies the conditions $|f(z)| < 1$, $f(0) = 0$, then
$$|f(z)| \leq |z| \text{ and } |f'(0)| \leq 1 \qquad (6.5.1)$$
If $|f(z)| = |z|$ for some $z \neq 0$, or if $|f'(0)| = 1$, then $f(z) = cz$ with a constant of absolute value 1.

**Proof** Suppose that
$$f(z) = c_1 z + c_2 z^2 + \cdots \quad (|z| < 1)$$
Let
$$g(z) = \frac{f(z)}{z} = c_1 + c_2 z + \cdots, \quad (z \neq 0)$$
where $g(0) = c_1 = f'(0)$. Then $g(z)$ is analytic in the open disc $|z| < 1$. On the circle $|z| = r < 1$, we have
$$|g(z)| \leq \frac{1}{r}$$
We apply the maximum principle to the function $g(z)$, thus
$$|g(z)| \leq \frac{1}{r} \text{ for } |z| \leq r$$
Letting $r$ tend to 1 we find that $|g(z)| \leq 1$ for all $z$, and this is the assertion of the theorem.

If the equality holds at a single point, it means that $|g(z)|$ attains its maximum and hence, that $g(z)$ must reduce to a constant.

**Theorem 6.5.2 (Lemma of Schwarz)**  If $f(z)$ is analytic for $|z|<R$ and satisfies the conditions $|f(z)|\leqslant M<+\infty$, $f(0)=0$, then

$$|f(z)|\leqslant \frac{M}{R}|z| \text{ and } |f'(0)|\leqslant \frac{M}{R} \qquad (6.5.2)$$

If $|f(z)|=\frac{M}{R}|z|$ for some $z\neq 0$, or if $|f'(0)|=1$, then $f(z)=\frac{M}{R}e^{i\alpha}z$ where $\alpha$ is a complex number.

***Proof***  If $f(z)$ is known to satisfy the conditions of the theorem in a disk of radius $R$, the original form of the lemma of Schwarz can be applied to the function $f(Rz)$. As a result we obtain $|f(Rz)|\leqslant|z|$, which can be rewritten as $|f(z)|\leqslant \frac{z}{R}$.

Similarly, if the upper bound of the modulus is $M$ instead of 1, we apply the theorem to $\frac{f(z)}{M}$ or, in the more general case, to $\frac{f(Rz)}{M}$.

**Theorem 6.5.3 (Lemma of Schwarz)**  If $f(z)$ is analytic for $|z|<R$ and satisfies the conditions $|f(z)|\leqslant M$, $f(z_0)=w_0$, then

$$\left|\frac{M[f(z)-w_0]}{M^2-\overline{w_0}f(z)}\right|\leqslant\left|\frac{R(z-z_0)}{R^2-\overline{z_0}z}\right| \qquad (6.5.3)$$

***Proof***  Let $\xi=Tz$ be a linear transformation which maps $|z|<R$ onto $|\xi|<1$ with $z_0$ going into origin, and let $Sw$ be a linear transformation with $Sw_0=0$ which maps $|w|<M$ onto $|Sw|<1$. It is clear that the function $Sf(T^{-1}\xi)$ satisfies the hypothesis of the original form of the lemma of Schwarz. Hence we obtain

$$|Sf(T^{-1}\xi)|\leqslant|\xi|$$

or

$$|Sf(z)|\leqslant|Tz|$$

This completes the proof.

We will show that all proper simply connected regions in $C$ are equivalent to the open disk $D=\{z:|z|<1\}$, and hence are equivalent to one another.

**Theorem 6.5.4 (Riemann Mapping Theorem)**  Given any simply connected region $D$ which is not the whole plane, and a point $z_0\in D$, there exists a

## Chapter VI  Conformal Mappings

unique analytic function $f(z)$ in $D$, normalized by the conditions $f(z_0)=0$, $f'(z_0)>0$, such that $f(z)$ defines a one-to-one mapping of $D$ onto the disk $|w|<1$.

**Theorem 6.5.5**  Let $f(z)$ be a topological mapping of a region $D$ onto a region $D'$. If $\{z_n\}$ or $z(t)$ tends to the boundary of $D'$, then $\{f(z_n)\}$ or $\{f(z(t))\}$ tends to the boundary of $D'$.

**Theorem 6.5.6 (Boundary Theorem)**  Suppose that the boundary of a simply connected region $D$ contains a line segment $v$ as a one-side free boundary arc. Then the function $f(z)$ which maps $D$ onto the unit can be extended to a function which is analytic and one to one on $D \cup v$. The image of $v$ is an arc $v'$ on the unit circle.

**Theorem 6.5.7**  Among the simply connected regions there are only two equivalence classes; one consisting of $C$ alone and the other containing all the proper simply connected regions.

### Exercises

1. Show by use of (6.5.3), or directly, that $|f(z)| \leq 1$ for $|z| \leq 1$ implies
$$\frac{|f'(z)|}{[1-|f(z)|^2]} \leq \frac{1}{1-|z|^2}$$

2. If $f(z)$ is analytic and Im $f(z) \geq 0$ for Im $z > 0$, show that
$$\frac{|f(z)-f(z_0)|}{|f(z)-\overline{f(z_0)}|} \leq \frac{|z-z_0|}{|z-\overline{z_0}|}$$
and
$$\frac{|f'(z)|}{\text{Im } f(z)} \leq \frac{1}{y} \quad (z=x+iy)$$

3. Let $f(z)$ be analytic on the unit disc $D$, and assume that $|f(z)|<1$ on the disc. Prove that if there exist two distinct points $a$, $b$ in the disc which are fixed points, that is, $f(a)=a$ and $f(b)=b$, then $f(z)=z$.

4. Let $f(z): D \to D$ be a conformal mapping of the disc into itself. Prove that for all $a \in D$ we have
$$\frac{|f'(a)|}{1-|f(a)|^2} \leq \frac{1}{1-|a|^2}$$

5. Prove that an isolated singularity of $f(z)$ is removable as soon as either Re $f(z)$ or Im $f(z)$ is bounded above or below.

6. Derive corresponding inequalities if $f(z)$ maps $|z|<1$ into the upper half plane.

7. Prove by use of Schwarz's lemma that every one-to-one conformal mapping of a disk onto another (or a half plane) is given by a linear transformation.

8. Suppose $|f(z)|\leqslant 1$ for $|z|<1$ and $f(z)$ is analytic. By considering the function $g(z)$ defined by
$$g(z)=\frac{f(z)-a}{1-\bar{a}f(z)}$$
where $a=f(0)$, prove that
$$\frac{|f(0)|-|z|}{1-|f(0)||z|}\leqslant |f(z)|\leqslant \frac{|f(0)|+|z|}{1+|f(0)||z|}$$
for $|z|<1$.

9. Suppose that Re $f(z)\geqslant 0$ for all $z$ in domain $D$ and suppose that $f(z)$ is analytic.

(a) Show that Re $f(z)>0$ for all $z$ in $D$.

(b) By using an appropriate Mobius transformation, apply Schwarz's Lemma to prove that if $f(0)=1$ then
$$|f(z)|\leqslant \frac{1+|z|}{1-|z|}$$
for $|z|<1$. What can be said if $f(0)\neq 1$.

(c) Show that $f(z)$ also satisfies
$$|f(z)|\geqslant \frac{1+|z|}{1-|z|}$$

# Appendix

## Appendix 1　常用数学专业英语词汇读写指南

### 一、数的表达

Complex Numbers(复数)　　Pure Imaginary Numbers(虚数)
Real Numbers(实数)　　Rational Numbers(有理数)
Irrational Numbers(无理数)　　Integers(整数)
Fractions(分数)　　Natural Numbers(自然数)
Zero(零)　　Negatives of Natural Numbers(负整数)
Decimals(小数)
Recurring and terminating Decimals(循环小数和非循环小数)
The set of numbers consists of real numbers and pure imaginary numbers.
(在数的系统中，包括实数和虚数)
Integers include natural numbers, zero and negative integers.
(整数包括自然数、零和负整数)
Even and Odd Numbers(偶数和奇数)
Even numbers refer to those that can be divided exactly by 2 and whose unit numbers are 0, 2, 4, 6 and 8, and those that cannot be divided by 2 are odd numbers, whose unit numbers are 1, 3, 5, 7 and 9.
(偶数是指能被 2 整除的数，其个位数是 0，2，4，6，8；不能被 2 整除的数是奇数，其个位数是 1，3，5，7，9)
Those numbers below zero are negative integers.
(小于零的自然数是负整数)
Decimals can be divided into terminating decimals and recurring or repeating decimals. (小数可分为非循环小数和循环小数或重复小数)
Cardinal Numbers(基数词)
Ordinal Numbers(序数词)
How to Express Fractions(分数的表达)

A fraction consists of a numerator, a denominator and a fraction stroke. In English, we usually use a cardinal to express the numerator and an ordinal to express the denominator. Except that the numerator is 1, all the cardinals must be plural. (分数由分子、分母和分数线组成。在英语中，一般用基数词表示分子，序数词表示分母。除了分子为"1"的情况外，序数词都要用复数)

For example(如)：

$\frac{1}{2}$　a (one) half

$\frac{1}{21}$　one twenty-first, one over twenty-one

$\frac{1}{4}$　one-fourth, a (one) quarter　　　　$\frac{1}{3}$　one third

$\frac{2}{3}$　two-thirds, two over three, two divided by three

$5\frac{1}{3}$　five and a third

$\frac{3}{4}$　three-quarters, three fourths　　　　$2\frac{1}{4}$　two and a quarter

$30\frac{7}{8}$　thirty and seven-eighths　　　　$1\frac{1}{2}$　one and a half

$3\frac{4}{7}$　three and four over seven, three and four-sevenths

$5\frac{5}{8}$　five and five-eighths, five and five over eight, five and five divided by eight

## 二、符号的表达

### （一）基本运算符号

$+$　plus, positive

Example 2.1

$2+3$　two plus three　　　　$+7$　positive (plus) seven

$a+5$　$a$ plus five

$-$　minus, negative

# Appendix

Example 2.2

$8-4$   eight minus four       $10-b$   ten minus $b$

$-3\frac{3}{5}$   negative/minus three and three-fifths

$\pm$ ; $\mp$   plus or minus ; minus or plus

Example 2.3

$x=\pm 3$   $x$ equals plus or minus three

$y=\mp 5$   $y$ equals minus or plus five

$\times$   multiplied by, times

Example 2.4

$5\times 6$   five times six or five multiplied by six

$x \cdot y$   $x$ times $y$ or $x$ multiplied by $y$

$-$, $/$, $\div$   over, is to, divided by

Example 2.5

$\frac{3}{4}$   three over       $9\div 4$   nine divided by four

$s/t$   $s$ divided by $t$

## (二) 括号

(   )   round brackets, parentheses

Example 2.6   $\left(6+5\frac{3}{4}-4\times 3\right)\div 3\frac{1}{7}$

Round brackets (Open parenthesis) six plus five and three-fourths minus four multiplied by three brackets closed (close parenthesis) divided by three and one seventh

Example 2.7   $\dfrac{x^3+1}{(x^2+1)^{\frac{2}{3}}}$

$x$ to the third power plus one divided by the quantity $x$ squared plus one to the two-thirds power

[   ]   square brackets

Example 2.8   $15x^2-[-4x^2+6(x-x^2)-3x]$

fifteen times $x$ squared minus square bracket minus four times $x$ squared plus six open parenthesis $x$ minus $x$ squared close parenthesis minus three times $x$ close bracket

# Functions of Complex Variables

| | |
|---|---|
| { } | braces |

Example 2.9  $\frac{1}{2}\{a[b+(c-d)]\}$

one-half open brace, $a$ open bracket $b$ plus open parenthesis $c$ minus $d$ close parenthesis close bracket close brace

| | |
|---|---|
| $=$ | be equal to, equals |
| $\equiv$ | be identical with, be identical to, be equivalent to |
| $\neq$ | be not equal to |
| $\approx$ | be approximately equal to |
| $<\ >$ | be less than (小于); be greater than |
| $\leqslant\ \geqslant$ | be less than or equal to; be greater than or equal to |
| $\mid\ \mid$ | absolute value |
| $n!$ | factorial $n$ |
| $\Sigma$ | the sum of the terms indicated, summation of |
| $\Pi$ | the product of the terms indicated |
| max | maximum value |
| min | minimum value |
| $x^n$ | Power is written as $x^n$, expressing the $n$ th power of $x$ or $x$ to the $n$ th power. |
| $\sqrt[n]{x}$ | The extracting of root is written as $\sqrt[n]{x}$ or $x^{\frac{1}{n}}$, which expresses $n$ th root of $x$ or $x$ to the quantity one over $n$ power. |

## 三、简单函数与区间

### (一) 函数

Example 3.1  $f(x)=ax^2+bx+c,\ a\neq 0$

The function of $x$ equals $a$ times the square of $x$ plus $b$ times $x$ plus $c$, where $a$ is not equal to zero.

→ approach, tend to Example 3.2  $a_n \to a$

$a$ sub $n$ approaches/tends to the value of $a$

Example 3.3  $a_n \to \infty$

$a$ sub $n$ approaches/tends to infinity

Example 3.4  $\lim\limits_{n\to\infty}\frac{1}{n^2}(1+2+\cdots+n)$

# Appendix

The limit as $n$ approaches infinity of the quantity one over $n$ squared times the sum of $n$ positive integers.

or

The limit as $n$ approaches infinity of the quantity one over $n$ squared times One plus two up to plus $n$.

Example 3.5  $\lim\limits_{n\to\infty} S_n = \dfrac{1}{3}$

The limit of $S$ sub $n$ as $n$ gets arbitrarily large is one third.

Example 3.6  $\lim\limits_{x\to a} f(x)$

limit of the function of $x$ as $x$ approaches $a$

Example 3.7  $\lim\limits_{\substack{x\to a \\ y\to b}} f(x, y)$

limit of $f(x, y)$ (or the function of $x$ and $y$) as $x$ approaches $a$ and $y$ approaches $b$

$\lim\limits_{n\to\infty} \sup a_n$

upper limit of $a$ sub $n$ as $n$ approaches infinity

$\lim\limits_{n\to\infty} \inf a_n$

lower limit of $a$ sub $n$ as $n$ approaches infinity

## （二）区间

The open interval is denoted by $(a, b)$

The closed interval is denoted by $[a, b]$

The half-open or half-closed interval is written as $[a, b)$ or $(a, b]$

## （三）微分与积分

Derivatives（导数）

$f(x)$   function of $x$（$f(x)$的函数）

$\Delta x$   delta $x$ or the increment added to $x$

$\Delta f_x$   delta $f$ or the increment added to function of $x$

$\mathrm{d}f(x)$   the differential function $f$ of $x$

The first derivatives（一阶导数）

$f'(x)$   the first derivative of $f$ with respect to $x$

The second derivatives（二阶导数）

$f''(x)$   the second derivative of $f$ with respect to $x$

the $n$ th derivatives（$n$ 阶导数）

$f^{(n)}(x)$   the $n$ th derivative of $f$ with respect to $x$

partial derivatives（偏导数）

$f_x$ or $\dfrac{\partial f}{\partial x}$   the first partial derivative of $f$ with respect to $x$

$f_{xx}$ or $\dfrac{\partial^2 f}{\partial x^2}$   the second partial derivative of $f$ with respect to $x$

$f_{xy}$ or $\dfrac{\partial^2 f}{\partial x \partial y}$   the second mixed partial derivative of $f$ with respect to $x$ and $y$

complete differentials（全微分）

$$dz(x, y) = \frac{\partial z}{\partial x}dx + \frac{\partial z}{\partial y}dy$$

The differential of $z$ with respect to $x$ and $y$ equals the partial derivative of $z$ with respect to $x$ times the differential of $x$ plus the partial derivative of $z$ with respect to $y$ times the differential of $y$

Integral（积分）

indefinite integral（不定积分）

$$\int f(x)dx$$

The indefinite integral of a function $f$ with respect to $x$

Example 3.8   $\int (2x+3)dx$

The indefinite integral of the quantity two times $x$ plus three with respect to $x$

definite integral（定积分）

$$\int_a^b f(x)dx$$

The definite integral of the function $f$ with limits $a$ and $b$ (from $a$ to $b$ or from $x=a$ to $x=b$) with respect to $x$

Example 3.9   $\int_0^{\frac{\pi}{2}} \dfrac{1}{1+a\cos x}dx$

The integral from zero to pi over two of the quantity one over one plus $a$ times cosine of $x$ with respect to $x$

## 四、初等函数

Exponential Functions（指数函数）
$e^x$ means the exponential function of $x$ with base e
Logarithmic functions（对数函数）
Trigonometric Functions（三角函数）

| | | | | | | |
|---|---|---|---|---|---|---|
| sin | （sine） | 正弦 | | cos | （cosine） | 余弦 |
| tan | （tangent） | 正切 | | cot | （cotangent） | 余切 |
| sec | （secant） | 正割 | | csc | （cosecant） | 余割 |
| arcsin | （the inverse sine function） | | | | | 反正弦 |
| arccos | （the inverse cosine function） | | | | | 反余弦 |
| arctan | （the inverse tangent function） | | | | | 反正切 |
| arccot | （the inverse cotangent function） | | | | | 反余切 |

## 五、与英文原版有关的英语惯用读法

### 第一章　Complex Variables（复数）

sums and products（加法与乘积）
the algebraic and geometric structure（代数和几何结构）
the complex number system（复数系）
the real number system（实数系）
ordered pairs（有序数对）
the complex plane（复平面）
rectangular coordinates（直角坐标）
the set of complex numbers（复数集合）
the real axis（实轴）
the imaginary axis（虚轴）
pure imaginary numbers（纯虚数）
imaginary number（虚数）
the real and imaginary parts（实部和虚部）
basic algebraic properties（基本代数性质）
the commutative laws（交换律）
the associative laws（结合律）

## Functions of Complex Variables

the associative laws（分配律）
additive identity（加法单位元）
multiplicative identity（乘法单位元）
additive inverse（加法逆元）
multiplicative inverse（乘法逆元）
further properties（进一步的性质）
the binomial formula（二项式公式）
liner simultaneous equations（线性联立方程）
mathematical induction（数学归纳法）
the cancellation law（消去律）
directed line segment（有向线段）
triangle inequality（三角不等式）
complex conjugates（共轭复数）
the conjugate of the sum is the sum of the conjugates（和的共轭等于共轭的和）
the conjugate of the difference is the difference of the conjugates（差的共轭等于共轭的差）
the conjugate of the product is the product of the conjugates（乘积的共轭等于共轭的乘积）
the conjugate of the quotient is the quotient of the conjugates（商的共轭等于共轭的商）
the generalized form（广义形式）
quadratic factor（二次因子）
exponential form（指数形式）
polar coordinates（极坐标）
in polar form（极形式，极型）
the principal value（主值）
parametric representation（参数表示）
products and quotients in exponential form（指数形式的乘积与商）
roots of complex numbers（复数的根）
de Moivre's formula（棣莫佛公式）
geometric argument（几何证明）
trigonometric identities（三角恒等式）

Chebyshev polynomials（切比雪夫多项式）
approximation theory（逼近论）
regular polygon（正多边形，正多角形）
the roots lies at the vertices of a regular polygon of $n$ sides inscribed in the circle（根位于圆内接 $n$ 边形的 $n$ 个顶点上）
the principal root（主值根）
equilateral triangle（等边三角形）
trigonometric identities（三角恒等式）
quadratic factor（二次因子）
quadratic equation（二次方程）
regions in the complex plane（复平面中的区域）
a deleted neighborhood（去心邻域）
interior point（内点）
exterior point（外点）
boundary point（边界点）
a set is open（开集）
a set is closed（闭集）
the closure of a set（闭包）
the punctured disk（去心圆盘）
a set is connected（连通集）
a polygonal line（折线）
a set is bounded（有界集）
accumulation point（聚点）

## 第二章　Analytic Functions（解析函数）

functions of a complex variable（解析函数）
a set of complex numbers（复数集合）
rule（法则）
domain of definition（定义域）
a real-valued function（实值函数）
a polynomial of degree $n$（$n$ 次多项式）
rational function（有理函数）
single-valued function（单值函数）

## Functions of Complex Variables

multiple-valued function（多值函数）
geometric characteristics（几何特征）
in the counterclockwise direction（逆时针方向）
mappings by the exponential function（指数函数映射）
elementary function（初等函数）
vertical and horizontal lines（水平线和垂直线）
for each positive number（对每一个正实数）
unique（唯一）
for any positive number（对任意正实数）
the smaller of the two numbers（两数中较小的一个）
nonnegative constant（非负常数）
arbitrarily small（任意小）
approach the origin along the real axis（沿实轴趋于原点）
approach along the imaginary axis（沿虚轴趋于）
theorems on limits（极限定理）
limits of functions of a complex variable（单复变函数的极限）
limits of real-valued of two real variables（二元变量的实函数的极限）
the definition of the limit（极限的定义）
limits involving the points at infinity（涉及无穷远点的极限）
the extended complex plane（扩充复平面）
the Riemann sphere（黎曼球面）
stereographic projection（球极投影）
the upper hemisphere（上半球面）
the finite plane（有限的平面）
maximum value（最大值）
minimum value（最小值）
complex constant（复常数）
the new complex variable（新的复变量）
throughout a neighborhood（整个邻域）
sufficiently small（充分小）
the definition of derivative（导数定义）
in any manner（任意的方式）

continuous partial derivatives of all orders（任意阶的连续偏导数）
differentiation formulas（微分公式）
positive integer（正整数）
negative integer（负整数）
chain rule（链式法则）
composite function（复合函数）
composite function（复合函数）
Cauchy-Riemann equations（柯西—黎曼方程）
a pair of equations（一对方程）
the first-order partial derivatives（一阶偏导数）
the first-order partial derivatives with respect to x of the function u（函数 $u$ 对于 $x$ 的一阶偏导数）
the right-hand side（右端）
evaluated（被赋值），evaluation（赋值，计算）
nonzero point（非零点）
sufficient conditions for differentiability（可微的充分条件）
mapping property（映射性质）
polar coordinates（极坐标）
straight forward to show（直接证明）
an alternative form（另一种形式）
the right-hand side（右边）
a fixed real number（定实数）
power functions（幂函数）
analytic functions（解析函数）
entire function（整函数）
singular point，singularity（奇点）
a necessary，but by no means sufficient，condition（必要但非充分条件）
sufficient condition（充分条件）
dot product（内积）
gradient vector（梯度向量）
harmonic functions（调和函数）
continuous partial derivatives of the first and second order（连续的一阶和二阶偏导数）

Laplace's equation（拉普拉斯方程）
electrostatic potential（静电势能）
three-dimensional space（三维空间）
free of charges（与电荷无关）
the semi-infinite vertical strip（无穷半带形）
steady temperature（定温度）
harmonic conjugate（共轭调和）
uniquely determined analytic functions（唯一确定的解析函数）
connected open set（连通开集）
a finite sequence of neighborhoods（一列有限邻域）

## 第三章　Elementary Functions（初等函数）

the exponential function（指数函数）
the complex exponential function（复指数函数）
with a pure imaginary period（具有纯虚数周期）
not expected（未知的）
the logarithmic function（对数函数）
nonzero complex number（非零复数）
single-valued（单值的）
negative real number（负实数）
branches and derivatives of logarithms（对数函数的导数和分支）
the multiple-valued logarithmic function（多值对数函数）
a branch of a multiple-valued function（多值函数的分支）
a branch cut（支割线）
a branch point（支点）
some identities involving Logarithms（一些关于对数的等式）
carry over to complex analysis（推广到复分析）
has $n$ distinct values（有 $n$ 个不同的值）
the right-hand side of the equation（方程的右端）
complex exponents（复指数）
powers of $z$（$z$ 的幂）
the principal branch（主值支）
the exponential function with base $c$（以 $c$ 为底的指数函数）

Trigonometric Functions（三角函数）

linear combination（线性组合）

hyperbolic function（双曲函数）

are not bounded on the complex plane（在复平面上是无界的）

the absolute value（绝对值）

a zero of a given function（给定函数的零点）

## 第四章　Integrals（积分）

derivatives of functions $W(t)$（函数 $W(t)$ 的导数）

definite integrals of functions $W(t)$（函数 $W(t)$ 的定积分）

improper integral（非正常积分）

piecewise continuous（分段连续）

has one-sided limits（有单侧极限）

the right-hand limit（右侧极限）

the left-hand limit（左侧极限）

the fundamental theorem of calculus（微积分学基本定理）

antiderivative（原函数）

with only minor modifications（做稍微修改）

a simple arc，or a Jordan arc（简单弧）

it does not cross itself（它不自相交）

a simple closed arc，or a Jordan curve（简单闭曲线）

the polygonal line（折线）

the unit circle（单位圆）

oriented in the counterclockwise direction（沿逆时针方向）

is traversed in the clockwise direction（沿顺时针方向）

is traversed twice in the counterclockwise direction（沿逆时针方向绕两次）

the parametric representation（参数表示）

not unique（不唯一）

differentiable arc（可微弧）

arc length（弧长）

the length of $C$（$C$ 的长度）

the unit tangent vector（单位切向量）

piecewise smooth arc（分段光滑弧）

a simple closed contour（简单闭围道）
the Jordan curve theorem（若尔当曲线定理）
contour integrals（围道积分）
a line integral（线积分）
the limit of a sum（和的极限）
the value of the integral（积分值）
the existence of the integral（积分的存在性）
upper bounds for moduli of contour integrals（围道积分模的上界）
has a maximum value（有最大值）
piecewise continuous（分段连续）
independent of path（与路径无关）
around closed paths（沿闭路径）
has an antiderivative（有原函数）
Cauchy-Goursat theorem（柯西-古萨定理）
described in the positive sense（正向）
proof the theorem（柯西—古萨定理的证明）
simply and multiply connected domains（单连通区域和多连通区域）
simply connected domain（单连通区域）
multiply connected domain（单连通区域）
annular domain（环域）
described in the counterclockwise direction（逆时针方向）
the principle of deformation of path（路径变形原则）
Cauchy integral formula（柯西积分公式）
taken in the positive sense（正向）
It tells us that if a function $f$ is to be analytic within and on a simple closed contour $C$, then the values of $f$ interior to $C$ are completely determined by the values of $f$ on $C$.
（它告诉我们如果函数 $f$ 在围线 $C$ 的内部及边界解析，则 $f$ 在围线 $C$ 内部的值完全由 $f$ 在围线 $C$ 上的值决定）
derivatives of analytic functions（解析函数的导数）
E. Morera（莫勒拉）
derivatives of all orders（任意阶导数）
derivatives of the first order（一阶导数）

derivatives of the second order（二阶导数）
have continuous partial derivatives of all orders（有任意阶连续偏导数）
Liouville's theorem and the Fundamental Theorem of Algebra（刘维尔定理和代数学基本定理）
Cauchy's inequality（柯西不等式）
Liouville's theorem（刘维尔定理）
the fundamental theorem of algebra（代数学基本定理）
maximum modulus principle（最大模原理）
Gauss's mean value theorem（高斯平均值定理）
maximum value（最大值）
maximum modulus principle（最大模原理）
minimum value（最小值）
minimum modulus principle（最小模原理）
increasing function（增函数）

## 第五章　Series（级数）

convergence of sequences（收敛序列）
an infinite sequence of complex numbers（一列无穷复数序列）
for sufficiently large values of $n$（对充分大的 $n$）
convergence of series（收敛级数）
an infinite series of complex numbers（一列无穷复数级数）
partial sums（部分和）
absolutely convergent（绝对收敛）
power series（幂级数）
Taylor series（泰勒级数）
Taylor's theorem（泰勒定理）
Taylor series expansion（泰勒展开式）
Maclaurin series（麦克劳林级数）
term by term differentiation（逐项微分）
expand the function into a series involving powers of $z$
（把函数展开成含有 $z$ 的幂次的级数）
Laurent Series（罗朗级数）

annular domain（圆环域）

Laurent's theorem（罗朗定理）

Laurent series（罗朗级数）

series representation（级数表示）

the Laurent series representation（罗朗级数表示）

Absolute and Uniform Convergence of Power Series（幂级数的绝对收敛和一致收敛）

absolute convergence of power series（幂级数的绝对收敛）

uniform convergence of power series（幂级数的一致收敛）

the circle of convergence of series（级数的收敛圆）

## 第六章　Residues and Poles（留数和极点）

a singular point is said to be isolate（奇点称为孤立的）

the punctured disk（有孔的圆盘）

the deleted neighborhood（去心邻域）

Cauchy's residues theorem（柯西留数定理）

except for a finite number of singular points（除了有限个孤立奇点）

using a single residue（单个留数的应用）

the three types of isolated singular points（孤立奇点的三种类型）

the portion of the series（级数的部分）

the principal part（主要部分）

a pole of order $m$（$m$ 阶极点）

a simple pole（简单极点）

a removable singular point（可去奇点）

an essential singular point（本性奇点）

residues at poles（极点处的留数）

inasmuch as（因为，由于）

errors of analytic functions（解析函数的零点）

a zero of order $m$（$m$ 阶零点）

the zero of an analytic function are isolated（解析函数的零点是孤立的）

line segment（线段）

zeros and poles（零点和极点）

behavior of a function near isolated singulat point（孤立奇点处的性质）

## 第七章　Applications of Residues（留数的应用）

evaluation of improper integrals（广义积分的计算）
the semi-infinite interval（半无限区间）
the Cauchy principal value（柯西主值）
even function（偶函数）
even rational functions（偶有理函数）
no factors in common（无公因子）
the distinct zeros（互异根）
improper integrals from Fourier analysis（傅里叶分析中的广义积分）
indented paths（不规则路径）
definite integrals involving sines and cosines（含有正弦和余弦的定积分）
argument principle（辐角原理）

# Appendix 2　与复变函数论有关的数学专业英语词汇

## A

abac(算图)
abacus(算盘)
abelian integral(Abe 积分)
abscissa(横坐标)
abscissa axis(横轴)
absolute constant(绝对常数)
absolute continuity(绝对连续)
absolute convergence(绝对收敛)
absolute derivative(绝对导数)
absolute deviation(绝对偏差)
absolute differential(绝对微分)
absolute differential　calculus(绝对微分学)
absolute differential(绝对微分)
absolute summability(绝对可和性)
absolute term(绝对项)
absolute value(绝对值)
absolute value of a complex number(复数的绝对值)
absolute uniform convergence(绝对一致收敛)
absolutely continuous(绝对连续的)
absolutely continuous function(绝对连续函数)
absolutely convergent(绝对收敛的)
absolutely convergent power series(绝对收敛幂级数)
absolutely integrable(绝对可积的)
absolutely integrable function(绝对可积函数)
accent sign(撇[号])
accumulation(聚点)
accumulation point(聚点)
accumulation principle(聚点原理)
acnode(孤立点)

## Appendix

actuarial mathematics（保险(统计)数学）
actual infinity(实无穷)
acute angle（锐角）
acute triangle（锐角三角形）
acyclic graph（非循环双图形，无圈图）
acyclic model（非循环模型）
acyclic network（非循环网络）
addition（加法）
addition of fractions(分数的加法)
addition of series(级数的加法)
addition theorem（加法定理）
adaptation theory（适应理论）
additive（加性，加法[的]）
adaptive control（适应性控制）
adaptive feedback control（适应反馈控制）
adaptive optimization（适应最优化）
adaptive system（适应性系统，自适应系统）
addition formulas（加法公式）
additive norm（加性范数）
additive number theory（堆垒数论）
additive theory of numbers（堆垒数论）
additive operation（加法运算）
additive operator（加性算子）
additive sequence（可加序列）
additive space（加法空间）
additivity（加性，可加性）
additivity of probablity（概率的可加性）
adjacency（邻接）
adjacency matrix（邻接矩阵）
adjacent(邻接的)
adjacent angles(邻角)
adjacent dihedral angles(邻接二面角)
adjacent lines(邻线)

adjacent points(邻点)
adjacent sides(邻边)
adjoint(伴随的)
adjoint determinant(伴随行列式)
adjoint matrix(伴随矩阵)
adjoint mapping(伴随映射)
adjoint operator(伴随算子，共扼算子)
adjoint polynomial(伴随多项式)
adjoint problem(伴随问题，共扼问题)
adjoint transformation(伴随变换)
adjugate determinant(转置伴随行列式)
adjugate matrix(转置伴随矩阵)
affine(仿射)
affine differential geometry(仿射微分几何)
affine geometry(仿射几何[学])
affine geometry in the narrower sence(狭义仿射几何)
affine mapping(仿射映射)
algebraic topology(代数拓扑)
allmost all(几乎所有)
allmost certain(几乎必然)
allmost certainly converge(几乎必然收敛)
allmost certainly convergent(几乎必然的)
allmost everywhere(几乎处处)
allmost everywhere converge(几乎处处收敛)
allmost everywhere convergent(几乎处处收敛的)
allmost everywhere divergence(几乎处处发散)
allmost everywhere divergent(几乎处处发散的)
allmost everywhere finite(几乎处处有限的)
analysis(解析)
analysis method(解析法)
analysis situs(拓扑学)
analytic(解析的)
analytic automorphism(解析自同构)

analytic bundle（解析丛）
analytic closed curve（解析闭曲线）
analytic completion（解析完备）
analytic continuation（解析开拓）
analytic extension（解析开拓）
analytically continuable（可解析开拓）
analytically continued（解析开拓的）
analytic continuation in the wider sense（广义的解析开拓）
analytic cover（解析覆盖）
analytic curve(解析曲线)
analytic differential(解析微分)
analytic equivalence(解析等价)
analytic expressions(解析式)
analytic family(解析族)
analytic function（解析函数）
analytic function in the wider sence（广义解析函数）
analytic function of several variables（多变量解析函数）
analytic geometry（解析几何学）
analytic mapping(解析映射)
analytic method(解析法)
analytic neighbo[u]rhood(解析邻域)
analytic number theory(解析数论)
analytic parameter(解析参数，解析参变量)
analytic relation(解析关系)
analytic set(解析集[合])
analytic solution(解析解)
analytic space(解析空间)
analytic structure(解析结构)
analytic subset(解析子集[合])
analytic subspace(复子空间)
analytic surface(解析曲面)
analytic theory(解析理论)
analytic transformation(解析变换)

analytic transformation group(解析变换群)
analytic trigonometry(解析三角学)
analytic vector(解析向量)
analytical(分析的，解析的)
analytical dynamics(解析动力学)
analytical form(解析形式)
analytical method(解析的法)
analytical model(解析模型)
analytical smoothing(解析光滑的)
analytical solution(解析解)
analytical system(分析系统)
analytical triangle(解析三角形)
analytically complete（解析函数完备的）
analytically complete domain（解析函数完备域）
analytically complete family（解析函数完备族）
analytically complete space（解析函数完备空间）
analytically continuable(可解析开拓的)
analytically continued(解析开拓的)
analytically dependent(解析相关的)
analytically independent(解析无关的)
analytically normal（解析正规的）
analyticity（解析性）
angle of rotation（旋转角）
angle-preserving（保角的）
annulus（圆环）
antiderivative（原函数）
apostrophe(撇号)
application of optimization（最优化的应用）
application of mathematics（数学的应用）
applied functional analysis（应用泛函分析）
applied mathematics（应用数学）
applied probability（应用概率论）
applied statistics（应用统计学）

approximate continuity(近似连续)
approximate convergence(近似收敛)
approximate derivative(近似导数，近似微商)
approximate functional equation(近似函数方程)
approximate integration(近似积分)
area coordinates(重心坐标)
argument(辐角)
argument function(辐角函数)
argument of a complex number(复数的辐角)
argument of vector(向量辐角)
argument principle(辐角原理)
arithmetic(算术的，四则的)
arithmetic expression(算术表达式)
arithmetic operation(算术运算)
arithmetic point(小数点)
arithmetic mean(算术中项，等差中项，算术平均)
arithmetic mean-geometric mean inequality(算术平均—几何平均不等式)
arithmetic progression(算术级数)
arithmetical progression(算术级数)
arithmetic series(算术级数)
arithmetical series(算术级数)
arithmetically equivalent(算术等价的)
arithmetic-geometrical(算术几何级数，等差等比级数)
arc（弧）
asymptoto（渐进线）

# B

barycenter（重心）
barycentric coordinates（重心坐标）
biocybernetics（生物控制论）
bio-mathematics（生物数学）
biometics（生物测量学）
biostatistics（生物统计学）
boundary（边界）

boundary point（边界点）
boundary problem（边界问题）
boundary value problem（边值问题）
bounded（有界的）
bounded above（上有界的）
bounded below（下有界的）
bounded closed regions（有界闭区域）
bounded closed set（有界闭集）
bounded closure（有界闭包）
bounded continuous functions（有界连续函数）
bounded convergence（有界收敛）
bounded function（有界函数）
bounded functional（有界泛函）
bounded harmonic function（有界调和函数）
bounded holomorphic function（有界全纯函数）
bounded interval（有界区间）
bounded linear functional（有界线性泛函）
bounded linear operator（有界线性算子）
bounded linear topological space（有界拓扑线性空间）
bounded linear transformation（有界线性变换）
bounded manifold（有界流形）
bounded measure（有界测度）
bounded measurable function（有界可测函数）
bounded variation（有界变差）
bounded variation function（有界变差函数）
bundle（丛）
bundle along fibre（沿纤维的丛）
bundle homomorphism（丛同态）
bundle homotopy（丛同伦）
bundle isomorphism（丛同构）
bundle mapping（丛映射）

# C

calculus（微积分[学]）

calculus of differences（差分法）
calculus of fluxion（微积分）
calculus of several variables（多变量微积分）
calculus of variations（变分法，变分学）
cancellation law（消去率）
cancellation law of addition（加法消去率）
cancellation law of multiplication（乘法消去率）
Cauchy condition for convergence of a series（柯西级数收敛条件）
Cauchy criterion for convergence（柯西收敛判别准则）
Cauchy inequality（柯西不等式）
Cauchy integral formula（柯西积分公式）
Cauchy's integral formula（柯西积分公式）
Cauchy integral theorem（柯西积分定理）
Cauchy mean value theorem（柯西均值定理）
Cauchy sequence（柯西序列）
Cauchy-Riemann differential equation（柯西—黎曼微分方程）
Cauchy's form of the remainder for Taylor's theorem（泰勒定理的柯西余项）
Cauchy's kernel（柯西核）
Cauchy's principal value（柯西主值）
Cauchy-Schwarz inequality（柯西—许瓦尔兹不等式）
center（心）
center of gravity（重心）
center of sphere（球心）
central（中心的）
central of a sphere（球心）
characteristic equation（特征方程）
characteristic equation of a matrix（矩阵的特征方程）
characteristic polynomial of a matrix（矩阵的特征多项式）
circle-chain method（圆链法）
circumcircle（外接圆）
circumference（圆周）
circumscribed（外切的）

circumscribed circle（外接圆）
circumscribed polygon（外切多边形）
close（闭）
closed（闭的）
closed collection（闭集）
closed set（闭集）
closed curve（闭曲线）
closed curve arc（闭曲线弧）
closed domain（闭域）
closed region（闭域）
closed smooth arc（闭光滑弧）
closure（闭包）
cluster point（聚点）
cluster point of a sequence（序列的聚点）
coefficient（系数）
combinatorial mathematics（组合数学）
commutative law（交换律）
commutative law of addition（加法交换律）
commutative law of multiplication（乘法交换律）
compact（紧的）
complete locally convex space（完备局部凸空间）
completeness of real numbers（实数的完备性）
completeness of the system of real numbers（实数系的完备性）
component（分量，分枝）
composition（复合）
conformal（保形的，保角的）
conformal deformation（保形变形）
conformal mapping（保角映射，保形映射）
conformality（保形性，保角性）
conformally（保形地，保角地）
conjugate complex numbers（共轭复数）
conjugate complex quantity（共轭复数）
connected region（连通区域）

connected set（连通集）
constrained extreme value（约束极值）
constrained game（约束对策）
constrained maximization problem（约束最大化问题）
constrained minimization problem（约束最小化问题）
constrained optimizations（约束最优化）
constrained system（约束系统）
constraint（约束）
constraint condition（约束条件）
continuation（连续，延拓）
continuity（连续性）
continuity from above（上连续）
continuity from below（下连续）
continuity from the left（左连续）
continuity from the right（由连续）
continuous（连续的）
continuous arc（连续弧）
continuous complex-valued functions（连续复值函数）
continuous curve（连续曲线）
continuous extension（连续延拓）
continuous function（连续函数）
continuously differentiable（连续可微的）
continuously differentiable curve（连续可微的曲线）
continuously differentiable for $n$-times（$n$ 次连续可微的）
continuously differentiable function（连续可微函数）
contours（围道）
contrafunctional（反函数的）
control system（控制系统）
control theory（控制理论）
convention（约定）
converge（收敛）
converge uniformly（一致收敛）
convergence（收敛）

convergence criterion(收敛判别法)

convergence domain(收敛域)

convergence in measure(依测度收敛)

convergence in norm(依范数收敛)

convergence in probability(依概率收敛)

convergence principle(收敛原则)

convergence radius(收敛半径)

convergence region(收敛区域)

convergence sequence(收敛序列)

convergence uniform(一致收敛)

convergent(收敛的)

convergent sequence(收敛序列)

convergent sequence of numbers(收敛数列)

convergent series(收敛级数)

convergent subsequence(收敛子序列)

convex programming(凸规划)

convex programming problem(凸规划问题)

convex quadratic programming problem(凸二次规划问题)

convex space(凸空间)

coordinate(坐标)

coordinate axis(坐标轴)

coordinate planes(坐标平面)

coordinate system(坐标系)

countable(可数的)

curve(曲线)

# D

denominator(分母)

dense(稠密)

dense everywhere(处处稠密)

dense set(稠密集)

dense subset(稠密子集)

dense-in itself(自稠密的)

denseness of rational number(有理数的稠密性)

density（稠密性）
density of rational numbers（有理数的稠密性）
denumerable（可数的）
denumerable class（可数集）
denumberable set（可数集）
descent function（下降函数）
derivable（可导的）
derivate（导数）
derivation（求导）
derivation calculus（求导演算）
derivative（导数，微商）
derivative on the left（左导数）
derivative on the right（右导数）
derivatives of higher order（高阶导数）
descending series（递减列）
determinant（行列式）
determinant of a matrix（矩阵的行列式）
determinant of coefficient（系数行列式）
determinant rank（行列式秩）
determinantal expansion（行列式展开式）
develop（展开）
development of a function（函数的展开）
difference（差，差分）
difference equation（差分方程）
differentiability（可微性）
differentiable（可微的）
differentiable almost everywhere（几乎处处可微）
differentiable real valued function（可微实值函数）
differential（微分的）
differentiation（微分）
differential and integral calculus（微积分）
differential calculus（微分学）
differential equation（微分方程）

differential equation of first order（一阶微分方程）
differential equation of higher order（高阶微分方程）
differential geometry（微分几何学）
differential law（微分法则）
differential quotient（导数，微商）
differentiable（可微的）
dilation（开拓，扩张，膨胀，伸缩）
dilation transformation（膨胀变换）
disc（圆盘）
disconnected（不连通的）
disconnected set（不连通集）
disconnecting set（不连通集）
discontinuous（不连续）
discontinuity（间断性，不连续性）
discontinuity of the first kind（第一类间断性）
discontinuity of the second kind（第二类间断性）
discontinuity point（不连续点）
discontinuity point of the first kind（第一类间断点）
discontinuity point of the second kind（第二类间断点）
disk（圆盘）
distributive law（分配律）
diverges（发散）
domain（区域）
domain of convergence（收敛域）
domain of definition（定义域）
dynamic model（动态模型）
dynamic optimization（动态最优化）
dynamic programming（动态规划）
dynamic system（动力系统）

# E

econometrics（计量经济学）
econometry（计量经济学）
edge（边）

eigenvalue（特征值）

eigenvalue of a matrix（矩阵的特征值）

eigenvalue of maximum（最大特征值）

element（元素）

ellipse（椭圆）

equator（赤道）

equicontinuity（同等连续性，等度连续性）

equicontinuous（同等连续的，等度连续的）

equicontinuous function（同等连续函数，等度连续函数）

expansion（展开，展开式）

expansion in series（级数展开）

expansion into power series（幂级数展开）

expansion of a determinant（行列式展开）

exterior to（外部的）

extremum（极值）

extremum values（极值）

## F

field（域）

field of complex numbers（复数域）

field of definitions（定义域）

foci（焦点）

formula of integration（积分公式）

fractional（分数）

function of a real variable（单实变函数）

functional space（函数空间）

functional value（函数值）

function-theoretic（函数论的）

fuzzy sets（模糊集）

fuzzy theory（模糊理论）

fuzzy topology（模糊拓扑）

## G

game theory（对策论）

general term（通项）

generalized limit（广义极限）
generalized theorem（推广的定理）
geometric progression（几何级数，等比级数）
geometric series（几何级数，等比级数）
geometrical series（几何级数，等比级数）
geometrically（几何上）
gradient（斜率，梯度）
gradient vector（梯度向量）
graph（图形）

# H

harmonic（调和的）
harmonic analysis（调和分析）
harmonic conjugate（调和共轭的）
homogeneous（齐次的）
homotopic（同伦的）
homotopic mapping（同伦映射）
homotopy（同伦）
homotopy chain（同伦链）
hyperbola（双曲线）

# I

idempotent（幂等的）
idempotent matrix（幂等矩阵）
identical（恒等的）
identical equation（恒等方程）
identical formula（恒等公式）
identical mapping（恒等映射）
identity（恒等式）
identity mapping（恒等映射）
identical matrix（单位矩阵）
identity transformation（恒等变换）
iff（当且仅当）
image（像）
imaginary unit（虚数单位）

implicit（隐的，隐式）
improper integral（反常积分，广义积分）
inclination（倾斜角）
increasing function（增函数）
increasing sequence（增序列）
increment（增量）
increment function（增量函数）
inferior limit（下极限）
infimum (inf)（下确界）
infinite integral（无穷积分）
infinite interval（无穷区间）
infinite limits（无穷极限）
infinite series（无穷级数）
infinitely great（无穷大）
infinitely large quantity（无穷大量）
infinitely samll（无穷小的）
infinitesimal（无穷小的）
infinitesimal analysis（微积分）
infinitesimal calculus（微积分学）
infinity（无穷大）
inflection point（拐点）
insulate（绝缘，隔离）
integer linear programming（整数线性规划）
integer programming（整数规划）
integrability（可积性）
integrable（可积的）
integrable function（可积函数）
integral（积分的，整的）
integral property（积分性质）
integrating factor（积分因子）
integration by parts（分部积分）
intercept（截距，截段）
interior to（内部的）

intersection（交，相交）
intersection point（交点）
inverse function（反函数）
inverse image（逆像）
individual（个体的）
index（指标）
inscribed（内接）
italics（斜体）

# L

least common denominator（最小公分母）
least common denominator（最小公倍数）
least upper bound（最小上界）
least value（最小值）
left continuous（左连续的）
left continuous function（左连续函数）
leg（三角形的侧边）
legitimate（合理的）
limit（极限）
limit function（极限函数）
limit inferior（下极限）
limit of a sequence（序列的极限）
limit of a sum（和的极限）
limit of function（函数极限）
limit point（极限点）
limit superior（上极限）
limit value（极限值）
limited function（有界函数）
limited set（有界集）
limited variation（有界变差）
limiting function（极限函数）
limiting point（极限点）
limiting point of a sequence（序列的极限点）
line segment（线段）

local concept（局部概念）
local control（局部控制）
local continuity（局部连续）
local convergence（局部收敛）
local convex（局部凸）
local property（局部性质）
lower bound（下界）
lower boundary（下边界）
lower bounded（下方有界的）
lower limit（下限）

## M

management science（管理学）
manifold（流形）
mathematical analysis（数学分析）
mathematics of computation（计算数学）
maximum norm（最大模）
mappings（映射）
moduli（模）
monotone（单调的）
monotone convergence theorem（单调收敛定理）
monotone decreasing function（单调递减函数）
monotone increasing function（单调递增函数）
monotone function（单调函数）
monotonically decreasing sequence（单调递减序列）
monotonically increasing sequence（单调递增序列）
monotonicity（单调性）
monotonicity principle（单调性）
multiplication（乘法）
multiplicity of root（根的阶，根的重数）
multiplicity of zero（零点的阶，零点的重数）

## N

$n$-dimensional Euclidean space（$n$-维欧几里得空间）
necessary and sufficient condition（必要且充分条件，充要条件）

necessary condition（必要条件）
neighborhood（邻域）
nondense（疏的，无处稠密的）
nondense set（无处稠密集，疏集）
non-denumerable aggregate（不可数集）
non-denumerable set（不可数集）
non-uniform（非一致的）
non-uniform approach（非一致趋近）
non-uniform convergence（非一致收敛）
non-uniform convergent（非一致收敛的）
norm（范数）
normal（正规的，垂直的，正交的，正态的）
normal family（正规族）
normal family of functions（正规函数族）
numerator（分子）
numerical analysis（数值分析）

# O

odd function（奇函数）
odd number（奇数）
odd-even（奇偶）
odevity（奇偶性）
oddness（奇性）
operation（运算）
opsearch（运筹学）
optimal control（最优控制）
optimal solution（最优解）
optimization（最优化）
optimization method（最优化方法）
optimum solution（最优解）
optimum seeking method（优选法）
order of infinitesimals（无穷小的阶）
order of infinity（无穷大的阶）
order of pole（极点的阶，极点的重数）

order of zeros（零点的阶，零点的重数）
orientation（定方向的）

## P

pairity（奇偶性）
parentheses（圆括号）
parity（奇偶性）
parametric programming（参数规划）
partial derivative（偏导数，偏微商）
partial differential（偏微分）
partial differential coefficient（偏导数，偏微商）
partial differential equation（偏微分方程）
partial increment（偏差，偏增量）
partial differentiable（可偏微的）
partial differentiable function（可偏微函数）
partially differentiate（偏微分）
period（周期）
periodic function（周期函数）
plane geometry（平面几何学）
points（点）
point of intersection（交点）
point of tangency（切点）
point range（点列）
point set（点集）
point set topology（点集拓扑）
point set theory（点集论）
point singularity（奇点）
polygon（多边形，多角形）
prime number（素数）
principle of the point accumulation（聚点原理）
principle of uniform boundness（一致有界原理）
provided（假设）

## Q

quadrant（象限）

quadratic（二次的）

quadratic programming problem（二次规划问题）

quadratic root（平方根）

quantity（量）

quotient（商）

R

radial direction（径向）

radian（弧度）

radical（根号，根式）

radius（半径）

radius of a circle（圆的半径）

radius of a solid sphere（球体的半径）

radius of a sphere（球的半径）

radius of convergence（收敛半径）

radius vector（径向量）

rectangular（长方行）

range of the function $f$（$f$ 的值域）

range of function（函数的值域）

rank（秩）

reciprocal（倒数的，互反的，互逆的）

reconcilable（同伦的）

rectangle（长方形，矩形）

rectangular（长方形的，矩形的，直角的）

rectangular coordinates（直角坐标）

region（带边区域）

regular linear transformation（正规线性映射）

regular polygon（正多边形）

remainder（余项）

residues（留数）

rational approximation（有理逼近）

rational entire function（有理整函数）

rational function（有理函数）

rational number field（有理数域）

rational point（有理点）
rational root（有理根）
reflection（反射）
right angle（直角）
rotation（旋转）

## S

schlicht（单叶）
schlicht function（单叶函数）
sectionally continuous（分段连续）
sectionally smooth（分段光滑）
sector（扇形）
semi-circle（半圆）
semi-circumference（半圆周）
semi-sphere（半球）
sensitivity analysis（灵敏度分析）
sequence of measurable functions（可测函数序列）
sequence of linear functions（线性泛函序列）
sequence of number（数列）
series（级数）
series expansion（级数展开）
series of functions（函数级数）
series with functions terms（函数项级数）
series of nonnegative terms（非负项级数）
series of positive terms（正项级数）
series of variable terms（变项级数）
simple closed curve（简单闭曲线）
simple curve（简单曲线）
simple function（简单函数）
simply connected domain（单连通域）
simply connected region（单连通域）
single-valued analytic function（单值解析函数）
single-valued complex function（单值复函数）
sink（释放）

slope（斜率）
smooth（光滑）
smooth curve（光滑曲线）
source（吸收）
space geometry（空间几何学）
spiral（螺旋线）
stipulate（约定）
strictly increasing（严格递增）
strictly increasing function（严格递增函数）
strictly monotone（严格单调）
strictly monotone decreasing（严格单调递减）
strictly monotone increasing（严格单调递增）
strictly monotone function（严格单调函数）
strictly monotone increasing function（严格单调递增函数）
strip（带）
subdomain（子区域）
subscript（下标）
summation（求和）
summation of series（级数求和）
subtraction（减法）
supremum（sup）（上确界）
supremum norm（上确界范数）
supremum theorem（上确界定理）
symmetry（对称性）
symmetry principle（对称原理）

# T

term（项）
theory of controls（控制理论）
theory of functions（函数论）
theory of games（对策论）
theory of optimal control（最优控制理论）
thin plates（薄圆盘）
total derivative（全导数，全微商）

total differential（全微分）
total differentiation（全微分）
total differential equation（全微分方程）
transcendental function（超越函数）
transcendental meromorphic function（超越亚纯函数）
transcendental number（超越数）
transformation（变换）
translation（平移）
trigonometry（三角学）

## U

unbounded（无界）
uniform approximation（一致逼近）
uniform boundness（一致有界）
uniform bounds（一致有界）
uniform continuity（一致连续性）
uniform continuity of functions（一致连续函数）
uniform convergence（一致收敛）
uniform convergence of a series（级数的一致收敛）
uniformly continuous（一致连续的）
uniformly continuous function（一致连续函数）
uniformly convergent（一致收敛的）
uniformly convergent sequence（一致收敛序列）
uniformly convergent series（一致收敛级数）
union（并）
unique（唯一的）
unique continuation（唯一开拓）
unique continuation theorem（唯一开拓定理）
unique decomposition theorem（唯一分解定理）
unique existence theorem（唯一存在定理）

## V

verification（核验）
vertice（顶点）
variate（变量）

variate-difference method(微量差分法，变量差分法)
variation(变分，变差)
variation equation(变分方程)
variation principle(变分原理)
variation of function(函数的变差)
variational(变分的)
variational calculuss(变分学，变分法)
variational method(变分法)
variational principle(变分原理)
vector(向量)
vector addition(向量加法)
vector arithmetic(向量运算)
volum(体积，容积)
vector field(向量场)

# W

weak(弱的)
weak accumulation point(弱聚点，弱极限点)
weak compactness(弱紧性)
weak convergence(弱收敛)
weak derivatives(弱导数)
weak differentiablity(弱可微性)
weak differential operator(弱微分算子)
weak integrability(弱可积性)
weak normality(弱正规性)
weak differentiable(弱可微性)
weakly integrabe(弱可积的)
weakly limiting(弱极限)

# Z

zero of a function(函数的零点)
zero of a polynomial(多项式的零点)
zero of order 1(一阶零点，单零点)
zero of order $n$($n$ 阶零点)
zero point(零点)

zero point of $k$-th order($k$ 阶零点)
zero point theorem(零点定理)
zero set(零集)
zero root(零根)
zero solution(零解)

# Answers

## Chapter 1  Complex Numbers and Functions

### §1  Complex Numbers

2. (a) $x^4+y^4-6x^2y^2$, $4xy(x^2-y^2)$  (b) $\dfrac{x}{x^2+y^2}$, $-\dfrac{y}{x^2+y^2}$

   (c) $\dfrac{x^2+y^2-1}{(x+1)^2+y^2}$, $\dfrac{2y}{(x+1)^2+y^2}$

   (d) $\dfrac{x^2-y^2}{x^4+y^4}$, $-\dfrac{2xy}{x^4+y^4}$

3. (a) $-\dfrac{1}{10}-\dfrac{3}{10}i$  (b) $3+i$  (c) $-2\pi+\pi i$  (d) $-8-6i$

   (e) $\dfrac{1}{2}-\dfrac{1}{2}i$  (f) $\dfrac{3}{10}-\dfrac{1}{10}i$  (g) $\dfrac{3}{5}+\dfrac{4}{5}i$  (h) $\dfrac{1}{2}-\dfrac{1}{2}i$

5. (a) $-11-2i$  (b) $-\dfrac{3}{5}-\dfrac{4}{5}i$

   (c) $-\dfrac{33}{169}+\dfrac{56}{169}i$  (d) $2^{n+1}\cos\dfrac{n\pi}{4}i$, $n=0, \pm 1, \pm 2, \cdots$

6. (a) $-\dfrac{2}{5}$  (b) $-\dfrac{1}{2}$  (c) $-4$

11. Suggestion: Reduce this inequality to $(|x|-|y|)^2 \geq 0$

12. (a) $32+24i$, $40$  (b) $2-\dfrac{3}{2}i$, $\dfrac{5}{2}$

    (c) $i$, $1$  (d) $-8i$, $8$

32. (a) $-\dfrac{3}{4}\pi$  (b) $\pi$

33. (a) $\sqrt{2}\,e^{i\frac{\pi}{4}}$  (b) $\sqrt{3}\,e^{i\arctan\sqrt{2}}$  (c) $3e^{i\pi}$  (d) $4e^{i\frac{\pi}{2}}$

    (e) $\sqrt{3}\,e^{-i\arctan\sqrt{2}}$  (f) $5e^{-i\frac{\pi}{2}}$  (g) $\sqrt{2}\,e^{-i\frac{3\pi}{4}}$

34. (a) $-1$  (b) $\dfrac{3\sqrt{2}}{2}+\dfrac{3\sqrt{2}}{2}i$  (c) $\dfrac{\pi}{2}-\dfrac{\sqrt{3}}{2}\pi i$

(d) $-1$  (e) $-\dfrac{\sqrt{2}}{2}-\dfrac{\sqrt{2}}{2}i$

41. (a) $\pm(1+i)$  (b) $\pm\dfrac{\sqrt{3}-i}{\sqrt{2}}$

(i) $\pm\sqrt{2}(1+i)$, $\pm\sqrt{2}(1-i)$  (j) $\pm(\sqrt{3}-i)$, $\pm(1+\sqrt{3}i)$

## §2 Regions in the Complex Plane

1. (a)(b)(d)(e)(h)(i)(k) is domain.

## §3 Functions of a Complex Variables

1. (a) $z\neq\pm i$  (b) $z\neq 0$  (c) $x\neq 0$  (d) $\mathrm{Re}z\neq 0$
2. $(x^3-3xy^2+x+1)+i(3x^2y-y^3+y)$
3. $\left(r+\dfrac{1}{r}\right)\cos\theta+i\left(r-\dfrac{1}{r}\right)\sin\theta$
4. $\bar{z}^2+2iz$
9. Suggestion: Do this by letting nonzero points $z=(x, 0)$ and $z=(x, x)$ approach the origin. Note that it is not sufficient to simply consider points $z=(x, 0)$ and $z=(0, y)$.

# Chapter 2  Analytic Functions

## §2 Cauchy-Riemann Equations

2. (a) $f''(z)=0$  (b) $f''(z)=f(z)$  (c) $f(z)=6z$
3. (a) $f'(z)=-\dfrac{1}{z^2}$, $z\neq 0$  (b) $f'(x+ix)=2x$  (c) $f'(0)=0$
4. (a) $f'(z)=(3x^2-3y^2)+6xyi=3z^2$
   (b) $f'(z)=e^z(z+1)$
   (c) $f'(z) = \cos x \mathrm{ch} y - i\sin x \mathrm{sh} y$

## §3 Elementary Functions

1. (a) $z=\ln 2+(2n+1)\pi i$  $(n=0, \pm 1, \pm 2, \cdots)$
   (b) $z=\ln 2+\left(2n+\dfrac{1}{3}\right)$  $(n=0, \pm 1, \pm 2, \cdots)$
   (c) $z=\dfrac{1}{2}+n\pi i$  $(n=0, \pm 1, \pm 2, \cdots)$

15. $\text{Re}(e^{e^z}) = e^{e^x \cos y} \cos(e^x \sin y)$, $\text{Im}(e^{e^z}) = e^{e^x \cos y} \sin(e^x \sin y)$

21. $\left(\dfrac{\pi}{2} + 2n\pi\right) \pm 4i \, (n=0, \pm1, \pm2, \cdots)$

22. $2n\pi + i\,\text{ch}^{-1}2$, or $2n\pi \pm i \ln(2+\sqrt{3})\,(n=0, \pm1, \pm2, \cdots)$

## §4  Multi-Valuled Functions

5. $z = i$

9. (a) $e^{-\frac{\pi}{2}}$  (b) $-e^{2\pi^2}$  (c) $e^{\pi}[\cos(2\ln 2) + i\sin(2\ln 2)]$

10. (a) $e^{-2n\pi + i\ln 2}$  (b) $e^{-(2n+1)\pi i}$

13. $c$ is real numbers.

16. (a) $\left(n+\dfrac{1}{2}\right)\pi + i\dfrac{\ln 3}{2}$  (b) $\dfrac{2n\pi + \pi - \arctan 2}{2} + \dfrac{\ln 5}{4}i$

    (c) $(2n+1)\pi i$  (d) $n\pi i$

# Chapter 3  Complex Integration

## §1  The Concept of Contour Integrals

1. Suggestion: In each part of this exercise, use the corresponding property of integrals of real-valued functions of $t$, which is graphically evident.

3. (a) $-\dfrac{1}{2} - i\ln 4$  (b) $\dfrac{\sqrt{3}}{4} + i\dfrac{1}{4}$  (c) $\dfrac{1}{z}$

4. $\dfrac{1}{2} + \dfrac{1}{2}i$

5. $\pi r^2 i$

6. $-\dfrac{1}{3} + \dfrac{1}{3}i$

7. (a) 1  (b) 2  (c) 2

8. (a) $-4 + 2\pi i$  (b) $4 + 2\pi i$  (c) $4\pi i$

9. (a) 0  (b) 0

10. $4(e^\pi - 1)$

11. $e^3 - e^{-3}$

12. (a) $\dfrac{e^{3-4i} - 1}{2}$  (b) $\dfrac{e^{2i}}{2} - \dfrac{1}{2}$

13. $1 - \cos(1+i)$

# Answers

14. $\dfrac{35}{12}+\dfrac{2}{3}i$

15. $\dfrac{1}{2}-\dfrac{1}{2}i$, $1-i$

19. Suggestion: By observing that, of all the points on that line segment, the midpoint is the closest to the origin.

26. $-2\pi i R^2 p'(a)$

28. Suggestion: make use of the equations $z\bar{z}=r^2$ and $|dz|=-ir\dfrac{dz}{z}$.

## § 2  Cauchy-Goursat Theorem

3. (a) $\dfrac{1+i}{\pi}$ (b) $e+\dfrac{1}{e}$ (c) 0

4. (a) $\dfrac{\sqrt{2}}{2}\pi i$ (b) $\dfrac{\sqrt{2}}{2}\pi i$ (c) $\sqrt{2}\pi i$

5. $2\pi(-6+13i)$

6. (a) $2\pi i$ (b) 0 (c) $2\pi i$

7. (a) $2\pi$ (b) $\dfrac{\pi}{4}i$ (c) $-\dfrac{\pi}{2}i$ (d) 0 (e) $i\pi\sec^2(x_0/2)$

8. (a) $\pi/2$ (b) $\pi/16$

## § 3  Harmonic Functions

1. (a) $f(z)=(x^2+xy-y^2)+i(2xy+\dfrac{y^2}{2}+\dfrac{x^2}{2}+\dfrac{1}{2})$

   (b) $f(z)=e^x(x\cos y-y\sin x)+ie^x(x\sin y+y\cos y)$

   (c) $f(z)=-\dfrac{x}{x^2+y^2}+i\dfrac{y}{x^2+y^2}+\dfrac{1}{2}$

4. Suggestion: Observe that the function $f(z)=u(x,y)+iv(x,y)$ is analytic in $D$ if and only if $-if(z)$ is analytic there.

# Chapter 4  Series

## § 1  Basic Properties of Series

5. (a) converge to 1 (b) converge to 0
   (c) divergent (d) converge to 1

12. (a) converge but not absolutely  (b) converge absolutely
    (c) divergent  (d) converge but not absolutely
    (e) divergent  (f) converge but not absolutely

## §2  Power Series

1. (a) 1  (b) 2  (c) 0  (d) 1
   (e) $+\infty$  (f) 0  (g) 1  (h) 27

6. $\sum\limits_{n=1}^{\infty} \dfrac{z^n}{n^2}$

8. Suggestion: Use summation by parts.

## §3  Taylor Series

1. (a) $\dfrac{1}{az+b} = \sum\limits_{n=0}^{\infty} (-1)^n \dfrac{a^n}{b^{n+1}} z^n$, $|z| < \left|\dfrac{b}{a}\right|$

   (b) $\int_0^z \dfrac{\sin z}{z} dz = \sum\limits_{n=0}^{\infty} (-1)^n \dfrac{z^{2n+1}}{(2n+1)(2n+1)!}$, $|z| < +\infty$

   (c) $\int_0^z e^{z^2} dz = \sum\limits_{n=0}^{\infty} \dfrac{z^{2n+1}}{n!(2n+1)}$, $|z| < +\infty$

   (d) $\sin^2 z = \dfrac{1}{2} \sum\limits_{n=0}^{\infty} (-1)^{n+1} \dfrac{(2z)^{2n}}{(2n)!}$; $|z| < +\infty$

   (e) $\dfrac{1}{(1-z)^2} = \sum\limits_{n=1}^{\infty} n z^{n-1}$, $|z| < 1$

2. (a) $\sin z = \sin[1+(z-1)] = \cos 1 \sin(z-1) + \sin 1 \cos(z-1)$

   (b) $\dfrac{z-1}{z+1} = -\sum\limits_{n=1}^{\infty} \left(-\dfrac{1}{2}\right)^n (z-1)^n$, $|z-1| < 2$

   (c) $\dfrac{z}{z^2-2z+5} = \sum\limits_{n=0}^{\infty} (-1)^n \dfrac{(z-1)^{2n+1}}{4^{n+1}} + \sum\limits_{n=0}^{\infty} (-1)^n \dfrac{(z-1)^{2n}}{4^{n+1}}$, $|z-1| < 2$

   (d) $\dfrac{2z+3}{z+1} = \sum\limits_{n=0}^{\infty} (-1)^n \dfrac{(z-1)^{n+1}}{2^n} + \dfrac{5}{2} \sum\limits_{n=0}^{\infty} (-1)^n \dfrac{(z-1)^n}{2^n}$, $|z-1| < 2$

   (e) $\sqrt[3]{z} = \sqrt[3]{1}[1+(z-1)]^{\frac{1}{3}}$

3. $\sum\limits_{n=0}^{\infty} \dfrac{(-1)^n}{3^{2n+2}} z^{4n+1}$  ($|z| < \sqrt{3}$)

4. $\sin z = \sin\left[\dfrac{\pi}{2} + \left(z - \dfrac{\pi}{2}\right)\right] = \cos\left(z - \dfrac{\pi}{2}\right)$

# Answers

$$\cos z = \cos\left[\frac{\pi}{2} + \left(z - \frac{\pi}{2}\right)\right] = -\sin\left(z - \frac{\pi}{2}\right)$$

11. Suggestion: Multiply and divide each term by $1-z$, and do a partial fraction decomposition, getting a telescoping effect.

12. (a) $e^z \sin z = \sum_{n=0}^{\infty} \frac{z^n}{n!} \sum_{n=0}^{\infty} (-1)^n \frac{z^{2n+1}}{(2n+1)!} = z + z^2 + \frac{z^3}{3} + \cdots, |z| < +\infty$

(b) $\sin z \cos z = \sum_{n=0}^{\infty} (-1)^n \frac{z^{2n+1}}{(2n+1)!} \sum_{n=0}^{\infty} (-1)^n \frac{z^{2n}}{(2n)!} = z - \frac{2}{3}z^3 + \cdots,$ $|z| < +\infty$

(c) $\dfrac{e^z - 1}{z} = \dfrac{\sum_{n=0}^{\infty} \frac{z^n}{n!} - 1}{z} = \sum_{n=1}^{\infty} \frac{z^{n-1}}{n!} = 1 + \frac{1}{2}z + \frac{z^2}{6} + \frac{z^3}{24} + \cdots, |z| < +\infty$

(d) $\dfrac{e^z - \cos z}{z} = \dfrac{\sum_{n=0}^{\infty} \frac{z^n}{n!} - \sum_{n=0}^{\infty} (-1)^n \frac{z^{2n}}{(2n)!}}{z} = 1 + z + \frac{z^2}{6} + \cdots, |z| < +\infty$

(e) $\dfrac{\cos z}{\sin z} = \dfrac{\sum_{n=0}^{\infty}(-1)^n \frac{z^{2n}}{(2n)!}}{\sum_{n=0}^{\infty}(-1)^n \frac{z^{2n+1}}{(2n+1)!}} = \frac{1}{z} - \frac{1}{3}z - \frac{z^3}{45} + \cdots, |z| < +\infty$

(f) $\dfrac{\sin z}{\cos z} = \dfrac{\sum_{n=0}^{\infty}(-1)^n \frac{z^{2n+1}}{(2n+1)!}}{\sum_{n=0}^{\infty}(-1)^n \frac{z^{2n}}{(2n)!}} = z + \frac{1}{3}z^3 + \cdots, |z| < +\infty$

(g) $\dfrac{e^z}{\sin z} = \dfrac{\sum_{n=0}^{\infty} \frac{z^n}{n!}}{\sum_{n=0}^{\infty}(-1)^n \frac{z^{2n+1}}{(2n+1)!}} = \frac{1}{z} + 1 + \frac{2}{3}z + \frac{1}{3}z^2 + \frac{13}{90}z^3 + \cdots,$

$|z| < +\infty$

13. $\dfrac{z^2}{z-2} = -\dfrac{[1+(z-1)]^2}{1-(z-1)} = -1 - 3(z-1) - 4(z-1)^2 - 4(z-1)^3 + \cdots,$
$|z-1| < 1$

14. $\dfrac{z-2}{(z+3)(z+2)} = -\dfrac{1}{12} + \dfrac{19}{12^2}(z-1) - \dfrac{121}{12^3}(z-1)^2 + \dfrac{619}{12^4}(z-1)^3 + \cdots,$
$|z-1| < 3$

22. Suggestion: The $n$-th coefficient of this series is not $\dfrac{(-1)^n}{n}$

23. Suggestion: Find a summation formula which is the analogue of intergration by parts.

## §4  Laurent Series

1. (a) $\dfrac{z+1}{z^2(z-1)} = \dfrac{1}{z^2}\left(1-2\sum\limits_{n=0}^{\infty} z^n\right)$, $0<|z|<1$;

   $\dfrac{z+1}{z^2(z-1)} = \dfrac{1}{z^2}\left(1+\dfrac{2}{z}\sum\limits_{n=0}^{\infty}\dfrac{1}{z^n}\right)$, $1<|z|<+\infty$

   (b) $\dfrac{z^2-2z+5}{(z-2)(z^2+1)} = -\sum\limits_{n=1}^{\infty}\dfrac{z^n}{2^{n+1}} - \dfrac{2}{z^2}\sum\limits_{n=1}^{\infty}(-1)^n\dfrac{1}{z^{2n}}$, $1<|z|<2$

   (c) $\dfrac{e^z}{z(z^2+1)} = \dfrac{\sum\limits_{n=0}^{\infty}\dfrac{z^n}{n!}}{z(z+1)} = \dfrac{1}{z}+1-\dfrac{1}{2}z-z^2+\cdots$, $|z|<+\infty$

2. $1+\sum\limits_{n=1}^{\infty}\dfrac{(-1)^n}{(2n+1)!}\cdot\dfrac{1}{z^{4n}}$

3. $\dfrac{e^z}{(z+1)^2} = \dfrac{e^{-1}e^{z+1}}{(z+1)^2}$, $0<|z+1|<+\infty$

4. $\sum\limits_{n=1}^{\infty}\dfrac{(-1)^{n+1}}{z^n}$

5. $\sum\limits_{n=0}^{\infty}z^n + \dfrac{1}{z}+\dfrac{1}{z^2}$  $(0<|z|<1)$   $-\sum\limits_{n=3}^{\infty}\dfrac{1}{z^n}$  $(1<|z|<\infty)$

6. (a) $-1-2\sum\limits_{n=1}^{\infty}z^n$   $(|z|<1)$    (b) $1+2\sum\limits_{n=1}^{\infty}\dfrac{1}{z^n}$

7. $\dfrac{z}{(z-1)(z-3)} = \dfrac{1+(z-1)}{2}\left(\dfrac{1}{z-3}-\dfrac{1}{z-1}\right)$, $0<|z-1|<2$

8. $\sum\limits_{n=0}^{\infty}(-1)^{n+1}z^{2n+1}+\dfrac{1}{z}$  $(0<|z|<1)$   $\sum\limits_{n=1}^{\infty}\dfrac{(-1)^{n+1}}{z^{2n+1}}$  $(1<|z|<\infty)$

9. (a) $\dfrac{1}{(z^2+1)^2} = \sum\limits_{n=0}^{\infty}(-1)^n(n+1)\dfrac{(z-i)^{n-2}}{(2i)^{n+2}}$, $0<|z-i|<2$

   (b) $z^2 e^{\frac{1}{z}} = \sum\limits_{n=2}^{\infty}\dfrac{1}{(n+2)!}\dfrac{1}{z^n}$, $0<|z|<+\infty$

   (c) $e^{\frac{1}{1-z}} = \sum\limits_{n=0}^{\infty}(-1)^n\dfrac{1}{n!}\dfrac{1}{(z-1)^n}$, $0<|z-1|<+\infty$

# Answers

$$e^{\frac{1}{1-z}} = 1 - \frac{1}{z} - \frac{1}{2z^2} - \frac{1}{6z^3} + \frac{1}{24z^4} + \cdots, \quad 1 < |z| < +\infty$$

## §5 Zeros of an Analytic Functions and Uniquely Determined Analytic Functions

1. (a) 4    (b) 15
2. Suggestion: Use Cauchy's inequality to show that the second derivative $f''(z)$ is zero everywhere in the plane. Note that the constant $M_R$ in Cauchy's inequality is less than or equal to $A(|z_0|+R)$.
3. Suggestion: Apply Liouville's theorem to the function $g(z)=\exp[f(z)]$
4. Suggestion: Do this by applying the corresponding result for maximum values to the function $g(z)=\dfrac{1}{f(z)}$.
5. Suggestion: Apply Liouville's theorem to the function $g(z)=-if(z)$

## §6 The three Types of Isolated Singular Points at a Finite Point

1. (a) $z=0$ is a simple pole; $z=\pm 2i$ is a pole of order 2; $z=\infty$ is removable singular point.

   (b) $z=k\pi-\dfrac{\pi}{4}$ is a simple pole; $z=\infty$ is not isolated singular point.

   (c) $z=(2k+1)\pi i$ is a simple pole; $z=\infty$ is not isolated singular point.

   (d) $z=\pm\dfrac{\sqrt{2}}{2}(1-i)$ is a pole of order 3; $z=\infty$ is removable singular point.

   (e) $z=k\pi+\dfrac{\pi}{2}$ is a pole of order 2; $z=\infty$ is not isolated singular point.

   (f) $z=-i$ is an essential singular point; $z=\infty$ is removable singular point.

   (g) $z=0$ is removable singular point; $z=\infty$ is an essential singular point.

   (h) $z=2k\pi i$ is a simple pole; $z=\infty$ is not isolated singular point.

2. (a) $z=0$ is removable singular point; $z=2k\pi i$ is a simple pole; $z=\infty$ is not isolated singular point.

   (b) $z=0$, $z=\infty$ is an essential singular point.

   (c) $z=0$ is removable singular point; $z=\infty$ is an essential singular point.

(d) $z=0$, $z=\infty$ is an essential singular point.

(e) $z=1$ is an essential singular point; $z=2k\pi i$ is a simple pole; $z=\infty$ is not isolated singular point.

6. Suggestion: $f(z)$ and $e^{f(z)}$ cannot have a common pole. Then apply Picard's Theorem.

## Chapter 5  Calculus of Residues

### §1  Residues

1. (a) $\operatorname*{Res}_{z=1}\dfrac{z}{(z-1)(z+1)^2}=\dfrac{1}{4}$; $\operatorname*{Res}_{z=-1}\dfrac{z}{(z-1)(z+1)^2}=-\dfrac{1}{4}$

   (b) $\operatorname*{Res}_{z=n\pi}\dfrac{1}{\sin z}=(-1)^n$

   (c) $\operatorname*{Res}_{z=0}\dfrac{1-e^{2z}}{z^4}=-\dfrac{4}{3}$ 
   (d) $\operatorname*{Res}_{z=1}e^{\frac{1}{z-1}}=1$

   (e) $\operatorname*{Res}_{z=1}\dfrac{z^{2n}}{(z-1)^n}=\dfrac{(2n)!}{(n-1)!\,(n+1)!}$

   (f) $\operatorname*{Res}_{z=1}\dfrac{e^z}{z^2-1}=\dfrac{e}{2}$; $\operatorname*{Res}_{z=-1}\dfrac{e^z}{z^2-1}=-\dfrac{e^{-1}}{2}$

2. (a) 1    (b) $-\dfrac{1}{2}$    (c) 0    (d) $-\dfrac{1}{45}$    (e) $\dfrac{7}{6}$

4. (a) $-2\pi i$    (b) $-\dfrac{2\pi}{e}i$    (c) $\dfrac{\pi}{3}$    (d) $2\pi i$

5. (a) $-2\pi i$    (b) 0    (c) $2\pi i$

6. (a) $\displaystyle\int_{|z|=1}\dfrac{dz}{z\sin z}=0$    (b) $\dfrac{1}{2\pi i}\displaystyle\int_{|z|=2}\dfrac{e^{zi}}{1+z^2}dz=\sin t$

   (c) $\displaystyle\int_C\dfrac{dz}{(z-1)^2(z^2+1)}=-\dfrac{\pi}{2}i$

7. (a) $m=1$, $\operatorname*{Res}_{z=0}f(z)=-\dfrac{1}{2}$    (b) $m=3$, $\operatorname*{Res}_{z=0}f(z)=-\dfrac{4}{3}$

   (c) $m=2$, $\operatorname*{Res}_{z=1}f(z)=2e^2$

8. (a) 3    (b) $-\dfrac{3}{16}$    (c) $\pm\dfrac{i}{2\pi}$

9. (a) $\pi i$    (b) $6\pi i$

# Answers

10. (a) $\pi i/32$  (b) 0
11. $4\pi i$
12. (a) $9\pi i$   (b) $-3\pi i$   (c) $2\pi i$
17. Suggestion: $\operatorname*{Res}_{z=\infty} f(z) = \operatorname*{Res}_{z=0}\left[-\frac{1}{z^2}f\left(\frac{1}{z}\right)\right] = -\frac{1}{2\pi i}\int_\gamma f(z)\mathrm{d}z$

$$= -\sum_{k=1}^{m} \operatorname*{Res}_{z=z_k} f(z)$$

when $\gamma(t)=Re^{it}$, $0\leqslant t\leqslant 2\pi$, for sufficiently large $R$.

## §2  Applications of Residue

1. (a) $\dfrac{2\pi}{\sqrt{a^2-1}}$   (b) $4\pi$   (c) $\dfrac{2\pi}{\sqrt{a(a+1)}}$   (d) $\dfrac{2\pi}{\sqrt{3}}$

 (e) $\dfrac{2\pi a}{(a^2-b^2)^{\frac{3}{2}}}$   (f) $\dfrac{3\pi}{8}$   (g) $\dfrac{2\pi}{\sqrt{1-a^2}}$   (h) $\dfrac{2\pi}{1-a^2}$

 (i) $\dfrac{a^2\pi}{1-a^2}$   (j) $\dfrac{a\pi}{(\sqrt{a^2-1})^3}$

 (k) $\begin{cases}\pi i\cdots\cdots\cdots a>0\\ -\pi i\cdots\cdots\cdots a<0\end{cases}$   (l) $\dfrac{\pi}{\sqrt{2}}$   (m) $\dfrac{\pi}{\sqrt{5}}$

 (n) $\dfrac{(2a+1)\pi}{4(a^2+a)^{\frac{3}{2}}}$   (o) $\dfrac{\pi}{\sqrt{1+a^2}}$   (p) $\dfrac{(2n)!}{2^{2n}(n!)^2}\pi$

3. (a) $\dfrac{\pi}{6}$   (b) $\dfrac{\pi}{2a}$   (c) $\dfrac{\pi}{4}$   (d) $\dfrac{\pi}{200}$

 (e) $\dfrac{2\pi}{3}$   (f) $\dfrac{\sqrt{2}\pi}{2}$   (g) $\dfrac{4\pi}{5}\sin\dfrac{2\pi}{5}$   (h) $-\dfrac{\pi}{5}$

4. (a) $\dfrac{\pi}{24e^3}(3e^2-1)$   (b) $\dfrac{\pi}{2a^2}e^{-\frac{ma}{\sqrt{2}}}\sin\dfrac{ma}{\sqrt{2}}$

 (c) $\dfrac{\pi}{2}e^{-a}$   (d) $\dfrac{\pi}{4b^3}(1+ab)e^{-ab}$

 (e) $\pi e^{-a}$   (f) $-\dfrac{\pi}{e}\sin 2$

 (g) $\dfrac{\pi}{e}(\sin 2-\cos 2)$

5. (a) $\dfrac{\pi}{2}$   (b) $\dfrac{\pi}{2}$   (c) $\dfrac{\pi}{a}\sin a$

(d) $\dfrac{\pi}{2a^2}(1-e^{-a})$   (e) $\dfrac{\pi}{2}\left(1-\dfrac{3}{2e}\right)$   (f) $\dfrac{\pi}{2}(b-a)$

6. Suggestion: Try the path from 0 to $R$, then $R$ to $Re^{i\frac{2\pi}{n}}$, then back to 0.

## §3  Argument Principle

1. (a) $4\pi$            (b) $-2\pi$          (c) $8\pi$
3. (a) 4               (b) 0
4. (a) 3               (b) 2               (c) 5
5. (a) 4               (b) 87
8. 3
10. Suggestion: Sketch the imagines of the imaginary axis and apply the argument principle to a large half disk.

# Chapter 6  Conformal Mappings

## §1  Analytic Transformation

4. Suggestion: use polar coordinates.

## §2  Rational Functions

3. (a) 0, 1   (b) 0, 1   (c) 2   (d) 0, 1   (e) $\pm i$   (f) 3
12. Suggestion: Let $T$ and $S$ be two such linear fractional transformations. Then, after pointing out why $S^{-1}[T(z_k)\} = z_k$ ($k=1, 2, 3$), use the results in Exercises 5 and 6 to show that $S^{-1}[T(z)\} = z$ for all $z$. Thus show that $T(z) = S(z)$ for all $z$.
14. Suggestion: Write the equation $T^{-1}(z) = T(z)$ as
$$(a+d)[cz^2 + (d-a)z - b] = 0$$

## §3  Fractional Linear Transformations

2. $w = (3z+2i)/(iz+6)$
3. $w - i = \dfrac{2z}{z-1}$
4. $w = -1/z$
5. $w = \dfrac{(z-z_1)(z_2-z_3)}{(z-z_3)(z_2-z_1)}$

## §4  Elementary Conformal Mappings

2. $0<v<1$
4. $v>u$
5. $-1<u<1$, $v<0$

## §5  The Riemann Mapping Theorem

5. Suggestion: Apply a fractional linear transformation.

# Bibliography

马立新. 2004. 复变函数学习指导. 济南：山东大学出版社.

童丽萍，陈治业. 2000. 数、符号、公式、图形的英文表达. 南京：东南大学出版社.

钟玉泉. 2000. 复变函数论. 3版. 北京：高等教育出版社.

John B. Conway. 1978. Functions of One Complex Variable. Second Edition. Springer-Verlag.

James Ward Brown & Ruel V. Churchill. 2005. Complex Variables and Applications. Seventh Edition. McGraw-Hill College.

Lars V. Ahlfors. 1979. Complex Analysis. Third Edition. McGraw-Hill.

Serge Lang. 1999. Complex Analysis. Fourth Edition. Springer-Verlag.